# Ion Chromatography
# Applications

Author

## Robert E. Smith, Ph.D.

Associate Professor
School of Pharmacy
University of Missouri, Kansas City

**CRC Press, Inc.**
**Boca Raton, Florida**

CHEMISTRY

**Library of Congress Cataloging-in-Publication Data**

Smith, Robert E., 1951-
  Ion chromatography applications.

  Includes bibliographies and index.
  1. Ion exchange chromatography. I. Title.
QD79.C453S66      1988        543′.0893        87-11769
ISBN 0-8493-4967-2

Direct all inquiries to CRC Press, Inc., 2000 Corporate Blvd., N.W., Boca Raton, Florida, 33431.

© 1988 by CRC Press, Inc.

International Standard Book Number 0-8493-4967-2

Library of Congress Card Number 87-11769
Printed in the United States

# PREFACE

When ion chromatography was invented in 1978, it presented a significant advance in determining common inorganic anions such as chloride, nitrate, phosphate, and sulfate, in a variety of aqueous solutions. For the next 3 to 4 years, methods were developed to determine almost any inorganic anion with a pKa under seven. This was made possible by using a suppressor column to suppress the conductivity of the eluent and enhance the conductivity of the analytes. In addition, alkali and alkaline earth metals, ammonium, and some aliphatic amines could be determined by using cation chromatography and a cation suppressor. At least two books and several review articles have been written describing such aspects of Ion Chromatography. This book describes numerous advances that have been made since then. Today, ion chromatography implies much more than isocratic elution, suppressor columns, and conductivity detection. This is the first time that gradient ion chromatography, mobile phase ion chromatography, pulsed amperometric detection, post-column reaction, and other newer techniques have been covered in one review. Throughout the book, practical applications are emphasized, with the largest chapter devoted, exclusively, to industrial and biological applications. This book is intended to be both a quick reference to the analyst who is performing methods of development, and a guide to a more detailed understanding of the different separation and detection methods now available in the chromatography of ions.

**Robert E. Smith**

# THE AUTHOR

**Robert E. Smith, Ph.D.,** is an Associate Professor at the University of Missouri — Kansas City School of Pharmacy, and has been performing analytical methods of development, with special emphasis on ion chromatography.

Dr. Smith received his B.S. degree in Chemistry from the University of Missouri in 1973, and his Ph.D. in Biochemistry in 1978. He was a postdoctoral fellow at the Sinclair Comparative Medicine Research Farm in Columbia, Missouri, in 1979, and was awarded a research fellowship at the Federal Institute in Zurich, Switzerland, in 1980. Since then, he has been in Kansas City, with research interests in epoxy resin chemistry, electroplating, and analytical biochemistry. This work has resulted in over 30 scientific papers. His current efforts are in using ion chromatography to determine sugar phosphates, especially inositol phosphates, in the ischemic brain.

# ACKNOWLEDGMENT

I would like to gratefully acknowledge the support provided by Dionex in preparing this book. Dionex is the owner of many exclusive patents in the field of ion chromatography. All art work and figures in this book were kindly provided by Dionex. All ion chromatographic data reported in this book were obtained using equipment on which Dionex owns patents. I would especially like to thank the research scientists at Dionex, such as Rosanne Slingsby, Dennis Gillen, and Roy Rocklin for all the information that they provided.

*This book is dedicated to the one person most responsible for my education. This book is dedicated to my Mother.*

# TABLE OF CONTENTS

Chapter 1
Basic Principles and Theory................................................ 1
I.    Introduction....................................................... 1
II.   Advances Since the Original Invention ........................... 10
III.  Micromembrane Suppressors....................................... 13
IV.   Anion Detection without Chemically Suppressed Conductivity ..... 18
V.    Advances in Cation Analysis .................................... 19
VI.   Mobile-Phase Ion Chromatography ................................ 20
VII.  Other Detection Methods ........................................ 21
VIII. Basic Definitions in Ion Chromatography......................... 25
IX.   Principles of Gradient Elution ................................. 29
X.    Conductivity ................................................... 37
XI.   Column Care..................................................... 41
References................................................................ 44

Chapter 2
Conventional Methods...................................................... 47
I.    Conductivity Detection.......................................... 47
II.   Analysis of Common Anions ...................................... 48
III.  Ion Chromatography Exclusion (ICE).............................. 64
IV.   Cations......................................................... 71
References................................................................ 73

Chapter 3
Nonconventional Methods................................................... 75
I.    Post-Column Reaction with UV-Visible or Fluorescence Detection.. 76
II.   Electrochemical Detection ...................................... 86
III.  Mobile-phase Ion Chromatography (MPIC).......................... 95
IV.   Automation .................................................... 106
References............................................................... 112

Chapter 4
Applications............................................................. 115
I.    Life Sciences ................................................. 115
II.   Environmental Sciences......................................... 131
III.  Plating Solution Analysis ..................................... 141
IV.   Other Applications ............................................ 155
References............................................................... 162
Index ................................................................... 167

Chapter 1

## BASIC PRINCIPLES AND THEORY

### I. INTRODUCTION

In 1974, just 1 year before the invention of ion chromatography, Horvath[1] described ion exchange chromatography (IEC) as the most prominent branch of liquid chromatography despite the lack of a universal detector for ions. Even in 1974, there were numerous applications for IEC. These included the analysis of organics, inorganics, and biopolymers. One application of IEC even helped win a Nobel Prize. Ion chromatography is simply a spin-off from these techniques. It is a modern method for ion analysis. It, too, can solve a number of very important analytical problems.

For example, consider the application which was responsible for a Nobel Prize. This application is amino acid analysis.[2] Using two different ion exchange columns, biochemists were able to separate and detect the 20 amino acids found in proteins and enzymes. The basic amino acids, i.e., those that have a net positive charge at pH 8, were separated on a cation exchange column that is relatively short. The neutral and acidic amino acids were separated on a longer anion exchange column. In both cases, the amino acids were detected by a post-column reaction with ninhydrin. The amino acids react with ninhydrin to form a colored complex which is detected by a UV-visible detector. This became an invaluable technique in studying the chemical properties of biologically significant polypeptides.

Amino acid analysis is not the only well-known application of ion chromatography. Another application which is commercially very successful, is water softening. Water softeners in the home contain ion exchange resin. The calcium and magnesium ions that contribute to the water hardness bind to the cation exchange resin in the water softener. As a result, the total calcium plus magnesium is significantly lowered. In each application, ions (such as an amino acid or calcium) in an aqueous solution, will bind to the ion exchange material under proper conditions of pH and ionic strength. These same ions may be eluted off the ion exchange material under different conditions of pH and ionic strength, thus, water softeners can be regenerated periodically. Often the analyst is required to take advantage of this difference in binding affinities to separate a mixture of ions into their individual components. Environmental chemists may want to perform an ion survey of different geographical sites and require multicomponent analysis. An enzymologist may want to separate a mixture of amino acids to determine the amino acid composition of a polypeptide. Both could use the same technique, ion chromatography, to accomplish their goals.

Ion chromatography is a subset of the broader field of IEC, but the two terms (ion chromatography and IEC) are not synonomous. Certainly there are numerous ion exchange methods for purifying a single enzyme from a tissue homogenate. However, at the time of writing, ion chromatographic methods for doing this, have not been described. However, the field of ion chromatography is growing rapidly and has been incorporated into Environmental Protection Agency (EPA) methods for water analysis.[3] Applications range from amino acid analysis to the determination of hydroxymethanesulfonate in fog water in Bakersfield, Calif.[4] Ion chromatography is used to help make the paper on which this book is printed, the microprocessors in hand-held calculators, and even in the food and beverages that we eat and drink.

The field of ion chromatography had its inception in 1975 with the landmark paper by Small et al.[5] They described an invention that made possible the simultaneous determination of inorganic anions and cations. The novelities of the invention were the use of low-capacity pellicular resins and chemical suppression of the conductivity of the eluent. A patent for ion chromatography was awarded to Dow Chemical Co., under whose auspices the technique was first developed. This original technology was licensed to the Dionex Corp., which now owns patents and trademarks on numerous hardware used in modern ion chromatography. The bulk of the literature published at the time of writing, describes the use of columns and chromatography systems that have since been considerably upgraded. Because of the constant upgrades, many publications describe the use of different ''separator'' and ''suppressor'' columns. Not only conductivity detectors, but also UV-visible, fluorescence, and electro-chemical detectors are used; the terminology rapidly becomes confusing. It is difficult for the biochemist, electroplater, environmental chemist, or even for the brewmaster to evaluate the data that are published on ion chromatographic analysis of carbohydrates, chromic acid, acid rain, or beer. By understanding the basic principles and terms used in ion chromatography, the different scientists can properly assess the importance of this relatively new area in analytical instrumentation.

Ion exchange can be compared to the binding of a metal to ethylenediamine tetraacetic acid (EDTA). EDTA can be compared to a column containing ion exchange sites. Both EDTA and a cation exchange column will bind metals at the proper pH. Both do so by exchanging loosely bound hydrogen ions for metal ions. EDTA at pH 6 to 9 contains two –COOH groups that act as ''cation exchange sites'' in this analogy. One calcium ($Ca^{2+}$) will exchange with two $H^+$ on the –COOH groups as illustrated below, where EDTA is abbreviated as $H_2L^{2-}$.

$$Ca^{2+} + H_2L^{2-} \rightarrow CaL^{2-} + 2^{H+}$$

Similarly, one calcium will exchange with two $H^+$ on the cation exchange resin. The difference here is that the two $H^+$ on the cation exchange resin are fixed on a solid support. The solid support is the ion exchange resin, which in many cases is a sulfonated polystyrene.

One can likewise compare the concentration of EDTA to the ion exchange capacity of the resin. For example, if there are 0.01 mol (2.42 g) of disodium EDTA, it would be able to bind 0.01 mol of calcium (or many other divalent metals). In the 0.01 mol of EDTA there are 0.02 mol of exchangeable $H^+$. If one mole of $H^+$ is defined as one ion exchange equivalent, the 0.01 mol of EDTA would represent 0.02 ion exchange equivalents. Similarly, one can measure the number of exchangeable $H^+$ on a cation exchange resin. This number of exchangeable $H^+$ can be expressed in terms of ion exchange equivalents. Suppose, then, that it takes 0.20 mg of ion exchange resin to have 0.02 ion exchange equivalents. The ion exchange capacity of the resin would then be 0.2/0.02 or 0.1 meq/g of resin. In fact, many of the cation exchange resins used in 1975 had ion exchange capacities of about this value.

Of course, there are some serious oversimplifications in this analogy of EDTA. Very different mechanisms are involved in ion exchange and in chelation with EDTA. More relevant to this discussion though, is the fact that the cation exchange resins are not small molecules with high mobility in solution, instead, they are polymers, with very limited mobility. The ion exchange sites on these polymeric resins are fixed on the solid, polymeric surface.

In order to attain 0.1 meq of ion exchange sites per gram of resin requires that the resin have a large surface area. This was accomplished by using polymeric resins that were relatively porous. These relatively high-capacity, porous, polymeric resins make it possible to do preparative-scale ion exchange chromatography. This is still quite useful in purifying large amounts of (ionic) biopolymers. Such porous resins do have their drawbacks, however. They compress when pressurized with an analytical pump. This makes it extremely difficult to incorporate porous resins into an analytical liquid chromatograph. This alone limits their reproducibility. In addition, porous ion exchange resins expand and contract as the ionic strength of the eluent changes. Unfortunately, changing the ionic strength of the eluent is a very useful way to elute strongly retained ions. As a result, it would be ideal to have an ion exchange resin that could effectively separate the required ions and not be susceptible to compression under pressure and not change volume when the ionic strength changes. This was accomplished by the introduction of polymeric resins with much lower porosity and, therefore, much lower ion exchange capacity. The term "pellicular" is used to denote low porosity and only surface modification. Thus, the term "pellicular resin" is often used to describe the low porosity and surface sulfonation of ion exchange columns.

The development of these lower-capacity pellicular resins was instrumental in the success of ion chromatography as an analytical technique. The pellicular resins used to separate ions in the field of chemically suppressed ion chromatography are now used in ion separator columns. They are simply a specific example of an ion exchange resin.

The term "ion exchange" implies that the primary mechanism of separation is due to exchange of eluent ions on the column for analyte ions. The column is packed with a substrate that contains a permanent charge. For example, the substrate may be a copolymer of polystyrene/divinylbenzene. The fixed charge is obtained by reacting the substrate with hot sulfuric acid to produce a sulfonated polystyrene/divinylbenzene. The sulfonate ($-SO_3H$) groups are strongly acidic so that they exist as $-SO_3^-$ $H^+$. The substrate now has a fixed negative charge from the $-SO_3^-$. Cations will be attracted to this negative charge and will exchange with the $H^+$. Thus, the sulfonated polystyrene/divinylbenzene is a cation exchange resin. The polystyrene/divinylbenzene resins used in 1974 were porous. This caused problems with compressibility under pressure, along with contraction and expansion as the eluent changed, as mentioned earlier. In addition, the analyte ions could diffuse into the pores of the resin. This permitted interactions between the analyte ions and the polystyrene/divinylbenzene. As a result, the separation mechanism was (and still is) a combination of ion exchange and nonionic interactions with the polymeric resin.

The porosity of the resin required slow flow rates and caused significant band spreading. This is typified by a standard experiment in quantitative analysis. Copper was determined by passing the unknown through a porous cation exchange resin in the $H^+$ form (i.e., the resin had $-SO_3H$ attached to it). The $Cu^{2+}$ exchanged with two $H^+$ on the resin. The $H^+$ eluted off the column and was collected in a flask to be titrated with NaOH. Because the copper diffused into the pores of the high-capacity resin, significant "band spreading" occurred. If the copper was originally dissolved in 10 m$\ell$, it would spread into a "band" of about 50 m$\ell$. In addition, the porosity of the resin required that the eluent flow rate be limited. If one attempted to increase the flow rate by using a high-pressure pump, the porous resin would simply compress, severely restricting the flow of effluent. In addition, the porous resin would expand and contract as the ionic strength of the eluent changed. This made it difficult to perform different analyses with the same packed column. If a stronger, higher ionic strength eluent was required, the resin would contract, creating voids in the column. This would severely limit the reproducibility of the chromatography. In practice, the copper analysis described required a 1 to 2 m$\ell$/min flow rate and each determination lasted 1 hr.

Despite these limitations, porous ion exchange resins were quite popular. One of the most popular ion exchange methods in 1974 was amino acid analysis, which was developed by Moore and Stein.[2] Again, the porous resins used in amino acid analysis meant that the separation mechanism was a combination of ion exchange and nonionic interactions with the polystyrene/divinylbenzene substrate. Although methods did exist for separating acidic, neutral, and basic amino acids simultaneously on one column, it was very common to use two columns. Contaminants could interfere with the determination of basic amino acids on the single column method, encouraging the use of one column (the short column) for basic amino acids and a second column (the long column) for acidic amino acids.[6] The amino acids were detected by post-column reaction with a reagent such as ninhydrin that reacts to form a highly colored compound that absorbs visible light.

There are several ions, though, that have no inherent physical properties (except conductivity) which permit their detection at parts per million (ppm) levels. Such anions include fluoride, chloride, nitrate, bromide, phosphate, and sulfate. Each of these could be detected individually by other methods, such as reaction with a color-generating reagent (i.e., molybdate for phosphate determination) or ion-selective electrodes. However, no such detection scheme could be used for all these anions if they were present in the same solution. The only detectable physical property that they have in common is conductivity. Thus, it was desirable to use a conductivity detector. The eluents used in ion exchange chromatography had high conductivity, forcing the analyst to detect low levels of conductivity (i.e., 3 $\mu$S) in the presence of a high background (i.e., 1000 $\mu$S).

Many of these problems (band spreading, expanding and contracting of resin, and high eluent conductivity) were overcome in the landmark paper by Small et al.[5] In this first paper describing the invention of ion chromatography, these three major problems were overcome. By using pellicular resins having low ion exchange capacity instead of high-capacity porous ion exchange resins, analyte diffusion and, therefore, band spreading, was minimized. The porous resins have ion exchange capacities as high as 1 meq/g whereas pellicular resins have ion exchange capacities as low as 0.02 meq/g. As mentioned earlier, the ion exchange capacity of a resin is simply a measure of the number of ion exchange sites per gram of resin.

The nonporous, or pellicular, nature of the resins of today also means that there is no expansion or contraction when different eluents are used. The resin does not compress when eluent is pumped through the column, permitting faster, controlled flow rates with limited back pressure. This permits the use of reproducibly packed analytical columns that can be used with an analytical pump to produce accurate eluent flow rates as high as 4 m$\ell$/min. Different analyses using different eluent ionic strengths could be used without creating column voids which appear with more porous columns.

This was certainly important in the original invention of ion chromatography. Perhaps more significantly, though, a second column was introduced which suppressed the conductivity of the eluent and enhanced the conductivity of the analyte ions, enabling the use of a conductivity detector. This second column was originally called a "stripper column" because it stripped the eluent of its conductivity. This column has also been described as a post-column reactor, because it does react with the eluent and analyte ions.[7] By reacting with the eluent, it suppressed its conductivity. This effectively lowered the background conductivity so that a conductivity detector became practical. It also helped that the second column reacted with the analyte ions, converting them to a highly conductive form. Thus, this second column, or suppressor column, suppresses the conductivity of the eluent and enhances the conductivity of the analyte ions. As a result, the major problem plaguing analysts in 1974, the lack of a universal detector for ions, was overcome. The conductivity detector became the long-sought "universal" detector. It was useful only when the suppressor column was used. Thus, this form of ion chromatography is now known as "chemically suppressed ion chromatography".

Chemical suppression of conductivity can be best understood by describing the details of a real analytical situation. For example, in cation analysis, the eluent is HCl. The HCl elutes sodium, ammonium, and potassium, which then go to the second column. This stripper column was simply an anion exchange column in the $-$OH form. The anions in the sample and eluent exchanged with the $-$OH on the stripper column. Thus, the highly conductive HCl was converted to HOH or water, which has low conductivity. If the sample contained NaCl, $NH_4Cl$, and KCl, the Cl$^-$ would exchange with the $-$OH, producing NaOH, $NH_4OH$, and KOH, which have higher conductivity than the corresponding chloride salts. Any cation in the sample would be converted to its hydroxide. This meant that transition metals would be converted to their insoluble hydroxides. Even though they eluted off the cation separator column, the transition metals never reached the conductivity detector, since they were trapped as a precipitate on the stripper column. Transition metals are detected, instead, by post-column reaction with a colorimetric reagent, PAR. This will be discussed in more detail in Chapter 3 (Section III).

In anion analysis, the anion stripper was a cation exchange resin in the $-H$ form, i.e., it was a sulfonated polystyrene/divinylbenzene. The $-H$ on the $-SO_3H$ was exchanged for the cations in the sample and eluent. In the original paper by Small et al. NaOH was used as the eluent. Thus, the highly conductive NaOH was converted to water. If the sample contained NaF, NaCl, $NaNO_3$, $Na_2HPO_4$, $NaNO_3$, NaBr, and $Na_2SO_4$, they would be converted to HF, HCl, $HNO_3$, $H_3PO_4$, $HNO_3$, HBr, and $H_2SO_4$. The $H^+$ ion has the highest specific conductivity of any cation, so the acid forms of the analyte anions had an elevated conductivity after leaving the stripper column. Thus, both the cation and anion stripper columns suppressed the conductivity of the eluent and enhanced the conductivity of the analyte ions. For this reason, the stripper columns have been called "suppressor columns" ever since they became commercially available.

This invention was patented by Dow, licensed to Dionex, and given the name "ion chromatography". The pellicular ion exchange columns became known as "separator columns". The original anion exchange column became known as the anion separator one, or AS-1, and the original cation exchange column became the cation separator one, or CS-1. The stripper column used in anion analysis became the anion suppressor column, or ASC-1, and for cation analysis, the cation suppressor column became CSC-1. The first series of instruments called in chromatographs (the Model 10, 12, 14, and 16) contained an eluent delivery system with eluent reservoirs, valves for selecting eluents, and a pump to deliver eluent at a controlled flow rate. No great care was taken to ensure pulseless pumping, since the conductivity detector used is quite insensitive to small variations in pressure. The original ion chromatographs contained the eluent delivery system, separator, and suppressor columns, conductivity detectors, a load/inject valve, a valve for column selection, and a valve to control regeneration of the suppressor columns.

The suppressor columns required periodic regeneration because of the nature of the suppression reaction. A fresh suppressor column for anion analysis contained a highly sulfonated polystyrene/divinylbenzene resin. As $Na^+$ in the eluent exchanged with $H^+$ on the $-SO_3H$, the suppressor was gradually converted to $-SO^3Na$. After all, the $-SO_3H$ had been converted to $-SO_3Na$, and it was no longer capable of suppressing the conductivity of the eluent or enhancing the conductivity of the analyte ions. To overcome this, it was necessary to regenerate the suppressor by converting the $-SO_3Na$ back to $-SO_3H$. This was done by washing the suppressor with an acid solution such as 1 $N$ $H_2SO_4$ for 12 to 15 min. In normal full-time use, a fully regenerated suppressor column would last 7 to 8 hr. before requiring regeneration. A single suppressor column could be regenerated several hundred times with no loss of performance.

Suppression for cation analysis was quite analogous. The cation suppressor column contained a polystyrene/divinylbenzene-based resin that contained quatenary ammonium hydroxide ($-NR_3OH$) sites. As HCl in the eluent and anions on the sample (i.e., Cl in NaCl, $NH_4Cl$, and KCl) eluted off the CS-1, the $-OH$ of the $-NR_3OH$ was exchanged for the Cl. This is illustrated below:

Eluent

$$\text{Resin}-NR_3OH + HCl \rightarrow \text{Resin}-NR_3Cl + HOH$$

Analytes

Resin—$NR_3Oh$ + $NaCl$ → Resin—$NR_3Cl$ + $NaOH$

Resin—$NR_3OH$ + $NH_4Cl$ → Resin—$NR_3Cl$ + $NH_4OH$

Resin—$NR_3OH$ + $KCl$ → Resin—$NR_3Cl$ + $KOH$

Thus, the highly conductive eluent, HCl, is converted to the weakly conductive $H_2O$, and the moderately conductive NaCl, $NH_4Cl$, and KCl are converted to the highly conductive NaOH, $NH_4OH$, and KOH. At the same time, the suppressor column is gradually converted to the $-NR_3Cl$ form, rendering it temporarily useless. After all the $-NR_3OH$ sites are converted to $-NR_3Cl$, the suppressor column would need to be regenerated, or reconverted to the $-OH$ form. In the original cation suppressor columns, this was done by washing with 1 $N$ NaOH for 12 to 15 min.

Thus, suppressor columns have two common features that are important to understand. Firstly, they suppress the conductivity of the eluent and enhance the conductivity of the analyte ions and secondly, they tend to be converted to an inactive form and need to be regenerated. This need for regeneration will be discussed later when describing further advances in suppressor column technology.

By introducing small particle size, reproducible, pellicular ion separator columns, and post-column chemical suppressor columns, Small et al.[5] effectively established a powerful new analytical technique for rapid, simultaneous determination of common inorganic anions and cations. After it became commercially available, ion chromatography became widely accepted for determining inorganic anions. Applications include analysis of chemical plant processes, airborne particulates and aerosols, and water samples (wastewater, plant effluents, drinking water, and precipitation). Thus, ion chromatography rapidly gained a reputation as a method for determining inorganic anions in dilute aqueous solutions. Early publications in this field were numerous and informative.[8-17] Two conferences were held in 1978 and 1979 and became the subject of two volumes.[16] Eventually, the Environmental Protection Agency (EPA), specified the ion chromatographic procedure as the method of choice for chloride, phosphate, nitrate, and sulfate determinations in precipitation samples in a recent quality assurance manual,[18] and has been incorporated in the American Society of Testing and Materials (ASTM) procedure for analysis of rain water.[19] During the first 5 to 8 years after its introduction, the vast majority of ion chromatographic methods were developed for inorganic ions, usually using ion exchange columns and isocratic eluents for separation and chemically suppressed conductivity for detection.

Although inorganic ion determinations are still quite numerous (and increasing daily), ion chromatography has developed much further. Ions need not be separated on ion exchange columns using isocratic elution. Gradient elution has recently been described.[20,21] Specialty ion separator columns have been developed for amino acid and carbohydrate analysis. Separations based on Donnan exclusion and ion pair chromatography are well known.[22,23] Chemically suppressed conductivity is no longer the sole method of detection or even the method of choice in many cases, instead, electrochemical detectors, post-column reactors with UV-visible or fluorescence detection are used.[24,25]

Ion chromatography has recently been defined as simply "the chromatography of ions". This definition implies that a variety of separation and detection modes can be mixed and matched to optimize the analysis of any given sample. This may also imply that the originally developed separator and suppressor columns are obsolete. However, the publications in the field of ion chromatography quite often report the use of these columns. Certainly, the results published in such reports are quite valid and reproducible. All that newer separator and suppressor columns are capable of, is permitting faster determinations of more substances in a single analysis. They also enable some ions to be determined that were difficult to determine previously. The "obsolete" columns are still quite useful. If cared for properly, they can have a lifetime of several years, so undoubtedly many future publications will appear which will use these "obsolete" columns.

It is important for both the analyst and the applications chemist to understand the basic principles of ion chromatography. In this way, the methods published can be clearly interpreted. For example, an environmental chemist may wish to draw some conclusions about acid rain based on ion chromatographic analyses. It might be confusing to read some of the original reports that used the "standard" anion separator (the AS-1) and later publications that use other separators and suppressors. Most ions in rain samples are determined by chemically suppressed ion chromatography, but some by electrochemical detection. In evaluating the data, the environmental chemist should be able to distinguish between differences in experimental approaches to ion chromatography. Most such differences simply involve advances in hardware. One report may describe the use of an AS-1 column and 16 min are required to separate chloride, nitrate, phosphate, and sulfate, whereas another report may use an AS-4 column and only 7 min are required for the same analysis. The important issue is that both methods accurately determine the levels of pollution and can be used in evaluating environmental concerns.

Similarly, a biochemist studying the flow of carbons in photosynthesis may see one report that describes the separation of sugar phosphates using AS-1 and AS-2 columns.[26] Another report used as AS-5 column.[27] Both reports used an isocratic eluent, and neither used the suppressor column that is now recommended, the anion micromembrane suppressor. If this biochemist wanted to develop a method to determine subparts per million levels of sugar phosphates, an understanding of ion chromatography would help in selecting the best, most up-to-date hardware available.

Before the introduction of reproducible pellicular, low-capacity ion exchange separator columns, ion exchange columns were porous and had high ion exchange capacity. Not all the ion exchange substrates were based on a polystyrene/divinylbenzene resin. Some biochemical applications, such as enzyme purification, used cellulose or dextran substrates to which groups such as diethylaminoethane (for anion exchange) and carboxymethyl (for cation exchange) were attached. For separation of small ionic species, the polystyrene/divinylbenzene substrate was used. It took years of developmental work at Dow Chemical Co., for Stevens and Small to produce the advanced technology polystyrene/divinylbenzene-based ion chromatography separator columns. To understand the differences between ion chromatography separators and conventional ion exchange resins, it is important to better describe the conventional resins.

Conventional polystyrene/divinylbenzene resins were produced by suspension copolymerization of styrene and divinylbenzene. The divinylbenzene, being bifunctional, produces the cross links which give the polymer more rigidity. The percent cross link that is mentioned in the literature is simply the amount of divinylbenzene that is present in the original reaction mixture. Reaction conditions would be carefully controlled to produce the narrowest possible molecular weight and particle size distributions. After polymerization, the polystyrene/divinylbenzene becomes rigid, hydrophobic spheres. The next step, was to introduce ion exchange sites. For example, if the resin were reacted with hot (100°C) sulfuric acid for 10 to 15 min, the phenyl groups in the polystyrene/divinylbenzene would become sulfonated. The sulfonic acid ion exchange groups, having identical negative charges, repel each other, causing the once rigid polymer to swell. The hydrophilic nature of the sulfonic acid groups also meant that the resin would absorb considerable moisture and swell even more. Thus, the sulfonated polystyrene/divinylbenzene beads were usually preswollen with the desired eluent to minimize equilibration time. These conventional resins had ion exchange capacities

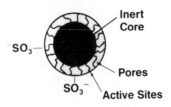

**Surface Sulfonated**

FIGURE 1.   IC separator resins — cation separator. Inert core is polystyrene/divinylbenzene.

of about 4.5 to 5.5 meq/g of dry resin. Because the resins were porous, ions could readily diffuse into the resin. This caused band spreading and limited column efficiency. Moreover, the high-capacity, gel-type resins would swell and contract considerably as the ionic strength of the eluent changed. Thus, it would have been impossible to use a step gradient such as 0.036 $M$ NaOH to 0.070 $M$ NaOH to 0.175 $M$ NaOH to 0.350 $M$ NaOH, similar to that used with a modern ion chromatography separator for amino acid analysis. Moreover, the high ion exchange capacities of conventional ion exchange resins required high ion fluxes for good separations. These ion fluxes would have been much too high to have been suppressed by the first suppressor columns. These problems were largely overcome by detailed developmental work at Dow.

One of the most detailed discussions of the earliest separator columns was written by Stevens and Small in 1978.[28] The cation separator was prepared by surface sulfonation of a styrene/divinylbenzene copolymer (2% divinylbenzene). This produced a surface shell of sulfonic acid groups. Thus, two distinctly different regions were formed, an inner, pellicular polystyrene/divinylbenzene core particle and an outer surface sulfonated shell. The column efficiency was found to depend on the percent cross linking (percent divinylbenzene) and the size of the polystyrene/divinylbenzene beads. First, they kept the bead size constant at 50 μm and varied the percent divinylbenzene from 0.04 to 12%. They obtained optimum chromatographic resolution using 2% divinylbenzene. They then experimentally determined the ion exchange capacity and used this value to calculate the depth of the sulfonation layer. These experiments and calculations of the depth of sulfonation were repeated for other bead sizes ranging from 40 to 400 μm. Their results indicated that a sulfonation depth of 175 to 200 Å was optimum. It was shown to be best to use this polystyrene/divinylbenzene that was only superficially surface-sulfonated. They also calculated the rate of diffusion into the pellicular ion separator columns they prepared. The rate was $1 \times 10^{-11}$ cm²/sec compared to $2 \times 10^{-7}$ cm²/sec for conventional ion exchangers.[29] This indicates that there is less diffusion of ions into the pellicular resin and much less band spreading. The band spreading decreases sensitivity and resolution so the columns were more efficient than the previously used porous resins.

The inert, pellicular polystyrene/divinylbenzene core provides a structural rigidity that permits much higher flow rates than those obtainable from conventional totally sulfonated ion exchangers. The polystyrene/divinylbenzene is chemically quite stable. The separator resins do not swell when the ionic strength of the eluent changes. There is even very little swelling when organic solvents such as acetonitrile or methanol are added to the eluent. Eluents ranging from pH 0 to 14, consisting of 3 $M$ nitric acid or 1.0 $M$ NaOH, can be used.

The cation separator is illustrated in Figure 1. The inert core is polystyrene/divinylbenzene. The active cation exchange sites ($-SO_3$) are located at the surface. There are also pores into which uncharged solutes can fit.

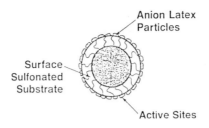

FIGURE 2. IC separator resins — anion separator. Active sites are quaternary amines.

The requirements of a good cation separator column as described by Pohl and Johnson[7] are as follows:

1. Low ion exchange capacity, i.e., 0.005 to 0.05 meq/g dry resin
2. Good pH stability
3. Wide range of selectivities

The low ion exchange capacity enables the use of low ion fluxes. The analyst is encouraged to analyze samples containing parts per million levels of cations and use relatively dilute eluents (i.e., 5 m$M$ HCl). Thus, the suppressor column can effectively convert all the highly conductive eluent to a low conductive form. The good pH stability, due to the inert polystyrene/divinylbenzene core, permits the use of acidic and basic eluents and the analysis of highly corrosive samples, such as those found in the electronics industry. The wide range of selectivities was not available at first, since the only cation separator available was the CS-1. However, other cation separators with different polystyrene/divinylbenzene particle sizes, different crosslinking, and different levels of sulfonation (ion exchange capacity) are now used.

Similar factors are involved in preparing anion separator columns. From the beginning (AS-1 column), ion chromatographers took advantage of the very strong electrostatic bonds formed between the negatively charged surface sulfonated polystyrene/divinylbenzene and positively charged aminated latex beads. The latex beads are much smaller than the polystyrene/divinylbenzene core particles. The positive charges on the latex beads are due to chemically bonded quaternary amino groups. As shown in Figure 2, this produced a resin with three distinctly different regions: a large inert core, a surface sulfonation layer, and an outer layer of positively charged (aminated) latex beads. This electrostatic bond between the aminated latex beads and the surface sulfonated polystyrene/divinylbenzene is so strong because of the large number of ionic interaction sites between each aminated latex particle and the sulfonated layer. This produces an essentially irreversible bond, which is stable to some organic solvents and extremes of pH. The latex beads, as shown in Figure 2, are much smaller than the polystyrene/divinylbenzene core particle, i.e., the latex beads are uniform in size, but vary from 0.1 to 0.5 μm (diameter) for different anion separator columns. There are small diffusion paths in the outer layers of an anion separator. The depth of the diffusion path is controlled by the sizes of the latex beads and the intermediate sulfonate layer. Thus, it is possible for substances such as detergents to be strongly retained on the core particle. More will be mentioned about the use of guard columns and other measures that can be taken to protect and clean separator columns.

## II. ADVANCES SINCE THE ORIGINAL INVENTION

One of the first developments was a separation technique called ion chromatography exclusion (ICE). Actually, ion exclusion itself was not a new development, but its combined use with chemically suppressed conductivity detection was new. One of the earliest papers on ion exclusion[22] described the first observations that when a porous ion exchange resin is placed in a dilute ionic solution, the concentration of the electrolyte is lower in the resin than in the surrounding solution. The ions are excluded from the resin, which preferentially absorbs water and any other uncharged substances present. As an example, they cited Dowex® 50 (a porous, sulfonated polystyrene/divinylbenzene) and its exclusion of HCl. The reason for this is that HCl is a strong acid (low pKa) and is fully dissociated into $H^+$ and $Cl^-$. The chloride is repelled by the $SO_3^-$ on the resin. This phenomenon of repulsion of like charges by a fixed charge is called "Donnan exclusion". The chloride cannot enter the pores of the ion exchange resin because of this Donnan exclusion. On the other hand, weak acids (high pKa), such as acetic and formic acid, would be only slightly dissociated in 1 m$M$ HCl. The fixed negative charges of the sulfonated resin does not repel the nearly uncharged carboxylic acids. Thus, a porous, high-capacity sulfonated resin can be used to separate weak acids from strong acids. Anions such as chloride, nitrate, bromide, and sulfate would be excluded and would elute in the void volume, whereas weaker acids such as phosphoric, acetic, and formic acids would be retained on the column and separated based on their pKa, or percent ionization. Thus, 20 μ$M$ phosphoric acid with a pKa of 2.12 would be 88% ionized in 1 m$M$ HCl, and 20 μ$M$ formic acid with a pKa of 3.75 would be 15% ionized and would be retained longer. It should be noted that any cations would also be strongly retained on the ICE separator column. In addition, benzoic acids and other aromatic compounds that can enter the resin pores are strongly retained because of their interaction with the polystyrene/divinylbenzene matrix. For detection, chemically suppressed conductivity is used. Until recently,[30] the suppressor column for ICE was a cation exchange resin in the silver form. The chloride in the HCl eluent will be precipitated onto the resin as AgCl. Thus, HCl is removed from the eluent:

$$Resin—Ag + HCl \rightarrow Resin—H + AgCl$$

Instead of regenerating this suppressor, the AgCl is removed by physically cutting the end off the column as it turns dark (usually twice a week). The suppressor column is packed with easily obtainable resin, so spent suppressors were discarded and easily refilled.

The next advances in anion chromatography were the development of new separator columns with different selectivities and/or enhanced resolution over the AS-1. By adjusting the percent crosslinking and size of the polystyrene/divinylbenzene core particle and latex beads, and the nature of the R group in the proprietary $-NR_3$ groups, column efficiency and selectivity could be controlled.

The second column developed by Dionex, the AS-2, had R groups that are more hydrophobic than those in the AS-1 column. The AS-2 column was first called the "brine column" because it was especially useful in determining trace anions in samples with high chloride content (see Chapter 2).

The next separator column, the AS-3, was called the "fast-run column" because it could separate the seven common anions in less time (10 vs. 20 min) than the AS-1 column using the same 3 m$M$ sodium bicarbonate per 2.4 m$M$ sodium carbonate eluent. This column, as with the AS-1 and AS-2, had approximated a 25-$\mu$m particle size and produced limited back-pressure, permitting the use of less precise pumps along with lower-pressure valves and fittings in the Model 10, 12, 14, and 16 ion chromatographs.

The next set of separator columns, however, used smaller polystyrene/divinylbenzene particle sizes which enhanced column efficiency, but created higher back-pressures. This required the introduction of the Series 2000 ion chromatographs. They have much more precise, pulseless analytical pumps, and higher-pressure fittings. The term "high-performance ion chromatography" (HPIC) was coined to describe this improved system with improved separator columns. Using the new anion separator, the AS-4, the standard seven anions eluted in the same order as they did on the AS-1 or AS-3, but they eluted faster on the AS-4. It provides better resolution and has more than twice the efficiency of the AS-1. The same bicarbonate/carbonate eluent was used for most applications with the AS-4. The AS-4, together with analytical pumps and a new suppressor column, were among the most significant advances in ion chromatography that had been seen since the original invention.

Another advance, this one in suppressor column technology, occurred at about this time. Originally, packed bed suppressors were used. They required periodic off-line regeneration, limiting the potential for automation. The advance accompanying the Series 2000 was the continuously regenerating fiber suppressor. As with the packed bed suppressor, the fiber suppressor column suppresses the conductivity of the eluent and enhances the conductivity of the analyte ions. The suppressor columns can be thought of as post-column reactors or stripper columns. They react with the eluent, stripping away the highly conductive components, and replaces them with lesser conductive components. Thus, the Na$^+$ in sodium bicarbonate and sodium carbonate (eluent for anion analysis) is stripped away and replaced by H$^+$ to form the low conductive carbonic acid. The chloride in HCl (eluent for cation analysis) is stripped away by the cation suppressor and is replaced with hydroxide to form water. In the process of suppressing the eluent conductivity, the suppressor also enhances the conductivity of the analyte ions. There is one drawback. In the process of suppressing the conductivity of the eluent, the exchange capacity, or stripping capacity, of the suppressor column gradually becomes exhausted. Anyone who owns a water softener is aware of this property of high-capacity ion exchange resins. In the case of the anion suppressor, all of the $-SO_3H$ is converted to $-NaSO_3$. Once this has happened, the packed bed anion suppressor must be regenerated. This is done by pumping 1 $N$ sulfuric acid through the suppressor for about 15 min, reconverting the $-NaSO_3$ to $-SO_3H$. The suppressor can then be re-used. This need for periodic suppressor regeneration was a nuisance. This inconvenience was what prompted the development of the fiber suppressors.

In the fiber suppressors, ion exchange fibers are used to perform the task of post-column stripping (suppression) that was originally performed by the ion exchange resin in the original packed bed suppressors. The exterior of the fiber is continuously bathed in a regenerating solution. The fiber is permeable to hydrogen ions in the sulfuric acid-regenerating solution (anion fiber suppressor). The sulfate in the sulfuric acid is excluded from the fiber by Donnan exclusion. However, Donnan exclusion is effective only at sulfuric acid concentrations below about 0.01 $M$. If regenerant concentrations above 0.01 $M$ are used, some of the sulfate would leak through, causing an elevated background conductivity. A small background conductivity is acceptable, so the upper concentration of the regenerant solution can be 0.01 to 0.1 $M$.

If a hollow fiber alone is used in the suppressor, laminar-flow could occur within the fiber. This would cause excess band broadening. To avoid this, the anion fiber suppressor is packed with neutral polystyrene/divinylbenzene. Thus, little band broadening occurs. The anion fiber suppressor, then, can be thought of as existing in a state of dynamic equilibrium. There are three distinct regions in an anion fiber suppressor:

1.    A totally expended region where eluent enters the fiber and the ion exchange fiber exists as $-NaSO_3$
2.    A totally regenerated region where regenerant solution enters and the ion exchange fiber exists as $-SO_3H$
3.    A region between these two extremes where part of the fiber is $-NaSO_3$ and the other is $-SO_3H$

The fiber is permeable to sodium and hydrogen ions. The eluent enters through one hole at the top of the column, and the Na of the sodium bicarbonate/sodium carbonate eluent is stripped away at the fiber membrane. The sulfuric acid regenerant enters at the bottom of the column. The $H^+$ migrates across the membrane and reconvertes $-NaSO_3$ to $-SO_3H$. The regenerant flow is fed by gravity, i.e., a large reservoir of regenerant is placed above the ion chromatograph. A valve can be used to control the flow of regenerant so that regenerant is not wasted when the ion chromatograph is being used.

There are three advantages of the anion fiber suppressor. First and probably most important, the ion chromatograph need not be shut-down periodically to regenerate a suppressor column. Second, weakly ionized ions such as nitrite are not subject to diffusion into pores as is the packed-bed suppressor. This diffusion caused band-spreading and loss of sensitivity. Since there are no large pores in the fiber suppressor, band-spreading for nitrite does not occur. Signal to noise ratio and reproducibility is improved. The third advantage is related to the phenomenon called the water "dip". In all ion chromatographs of samples prepared in deionized water, there is always a negative peak just ahead of the fluoride peak. With packed-bed suppressors, the retention time of the water dip varied as the suppressor column became expended. With the anion fiber suppressor, this water dip remained constant.

The anion fiber suppressor did have two limitations. It could not be used with a new type of separation that is called "mobile-phase ion chromatography" (MPIC). MPIC is analogous to reversed-phase HPLC, except that MPIC columns contain polystyrene/divinylbenzene and not long-chain alkyl groups (C-8 or C-18) chemically bonded to silica (as is common in reversed-phase HPLC). In addition, MPIC, in its beginning years, most often used chemically suppressed conductivity for detection, and not a UV detector as is used in HPLC. However, MPIC, like reversed-phase HPLC, utilizes eluents that contain acetonitrile, and often use a detergent as an ion pair reagent (see Chapter 3 for more details).

Although MPIC immediately proved its utility in separating large, hydrophobic anions, it could not use the original anion fiber suppressors. The original fiber suppressors (AFS-1 and CFS-1) are damaged by acetonitrile and ion-pair reagents such as tetrapropyl ammonium hydroxide. Thus, the first MPIC methods still used the packed-bed suppressor which was not damaged by acetonitrile or ion-pair reagents. The second was that only a limited ion flux was allowed. The anion fiber had a limited ion exchange capacity, so there was a limit to the concentration of eluent that could be suppressed. This meant that highly concentrated eluents could not be used and that gradient elution was not possible. The anion fiber suppressor was quite reliable, however. It did permit unattended automated analysis, because off-line regeneration was no longer needed.

After the development of the anion fiber suppressor, the next development was in separating large, hydrophobic anions. Several of these large, hydrophobic anions were strongly retained on the AS-4, however. This prompted the development of another separator column, the AS-5. Oxyanions, such as molybdate, tungstate, and chromate, could be eluted in reasonable times off the AS-5. The same bicarbonate/carbonate eluent could be used, but some of the later eluting peaks showed distinct peak tailing. This can be minimized by using the eluent recommended by Dionex, 8 m$M$ $p$-cyanophenol, 4.3 m$M$ sodium bicarbonate, 3.4 m$M$ sodium carbonate, plus 2% acetonitrile. The peak tailing is due to nonionic interactions with the polystyrene/divinylbenzene core resin and the large, hydrophobic anions. By adding $p$-cyanophenol and acetonitrile, this interaction is minimized because the $p$-cyanophenol and acetonitrile preferentially bind to the polystyrene/divinylbenzene. This effectively coats the polystyrene/divinylbenzene, blocking the interaction with the large anions.

The next separator column introduced was the AS-4A, which is quite similar to the AS-4, but it is much more rugged. One of the reasons for its development was because scientists analyzing soil samples found that the ill-defined humic acids would poison the AS-4 column, limiting its useful lifetime. The AS-4A was not poisoned by such samples. The AS-4A column also found good utility when a new, higher-capacity membrane suppressor became available. Much higher ion fluxes could be used, which popularized the use of eluents other than bicarbonate/carbonate.

Then came another very significant advance. Prior to February 1986, gradient elution was not possible within chromatography. Step gradients could be used with the Series 2000® ion chromatographs, but not continuous gradients. After the 1986 Pittsburgh Conference, however, gradient elution was introduced in the Series 4000® ion chromatographs. This advance was made possible not only by the development of a nonmetallic gradient pump, but also by the increased capacity of the new micromembrane suppressors.

## III. MICROMEMBRANE SUPPRESSORS

Prior to 1985, ion chromatographers were restricted to the use of packed-bed suppressors or fiber suppressors. As discussed previously, packed-bed suppressors required periodic off-line regeneration, and weak acid analytes (such as nitrite) could diffuse into the pores, causing band broadening. These two drawbacks were overcome by the introduction of the fiber suppressors. They suffered, however, from a limited ion exchange capacity. They cannot adequately suppress concentrated eluents that are needed for some applications. As a result, bicarbonate/carbonate became almost the only eluent used. Moreover, the anion fiber suppressor was not stable to acetonitrile and ion-pair reagents. This meant that some applications still required the use of packed-bed suppressors.

These problems were overcome by the introduction of the micromembrane suppressor (MMS). The MMS combines the high ion exchange capacity of the packed-bed suppressor with the constant regeneration capability of the fiber suppressors. The MMS has a void volume under 50 $\mu\ell$, which minimizes band spreading. Because of its higher ion exchange capacity, the MMS can suppress much higher ion fluxes. Thus, new eluents such as glycine, and more concentrated eluents, such as 95 m$M$ NaOH, can be used. More concentrated samples of NaOH and boric acid can also be analyzed without worrying so much about troublesome matrix effects. The MMS suppresses the conductivity of the eluent in much the same way as does the fiber suppressor. Sodium ions in the sodium bicarbonate/sodium carbonate eluent are exchanged for $H^+$ in the regenerant by means of a strong acid ion exchange membrane. The $Na^+$ exchanges with membrane-associated $H^+$ to form the membrane-Na. The regenerant then reconverts the membrane-H to membrane-Na. As with the fiber suppressor, the membrane suppressor is only semipermeable. It is permeable to cations ($Na^+$ and $H^+$), but excludes anions (carbonate, fluoride, chloride, nitrate, etc.). The eluent and regenerant flow in opposite directions, establishing a dynamic equilibrium similar to that described for fiber suppressors.

This dynamic equilibrium can be described in more precise terms. Dynamic suppression capacity has been defined as the maximum number of equivalents of eluent that can be neutralized per unit time.[20] This differs from packed bed suppressor capacity, which is the maximum equivalents of eluent that can be neutralized before exhausting all its ion exchange sites, and rendering it useless (until after regeneration). This was fixed by the capacity of the resin and the amount of resin in the column. Dynamic suppression capacity has units of microequivalents per minute. It is the product of eluent concentration in microequivalents per milliliter, and eluent flow rate in milliliters per minute. For example, if a suppressor can suppress 100 mN NaOH at a flow rate of 2 m$\ell$/min, the dynamic suppression capacity would be as follows:

$$(100 \ \mu eq/m\ell)(2.0 \ m\ell/min) = 200 \ \mu eq/min$$

Theoretically, dynamic suppression is independent of absolute concentration and flow rate. In other words, this same suppressor should be capable of suppressing 200 mN NaOH flowing at 1.0 m$\ell$/min. In practice, though, it is easier to suppress low concentrations at high flow rates than vice versa.

To understand the experimental parameters in using micromembrane suppressors it is important to know how they are designed. The suppressor is comprised of two sizes of ion-exchange screen. The fine screen is for the eluent chamber and minimizes the void-volume and dispersion of analytes. The coarse screen is for the regenerant chambers. It provides the necessary ion exchange capacity and flow charcteristics without restricting regenerant flow. The eluent screen is placed in the middle of the stack of screens. It is sandwiched on each side by ion-exchange membranes. On either side of each membrane is a regenerant screen. The stack is laminated together by pressure, causing intimate contact between screens and membranes.

Eluent passes through a hole in the upper regenerant screen and membrane. It enters the screen-filled eluent chamber and passes through the chamber. It then goes out through a second set of holes in the upper membrane and regenerant screen. Regenerant flows countercurrent to the eluent through the screen-filled regenerant chamber.

For cation analysis, the eluent is, in the simplest case, HCl. The suppressor has anion exchange sites. It exchanges hydroxide for chloride, converting HCl to water. Two mechanisms were described for this process of ion exchange. In the first, the hydroxide is simply eluted by the chloride in the eluent. The hydroxide combines with the remaining $H^+$ in the eluent to form water. The second mechanism is based on the fact that water is a much weaker base than the ion-exchange site. The first step in this mechanism involves the formation of water by reaction of the hydronium ion in the eluent, with hydroxide ion from a regenerated ion exchange site. This causes an excess negative charge in the eluent (chloride anions) and an excess positive charge in the membrane (unsatisfied cationic exchange sites). This creates an electrical field. Chloride ions then move under the force of this field toward the membrane. This fills the ion exchange site and neutralizes the charge imbalance.

Both mechanisms probably operate in a membrane suppressor. At the suppressor inlet, there is a high concentration of chloride ions, and statistically there is a higher probability that chloride will elute hydroxide from an exchange site. Near the exit of the suppressor there are significantly fewer chloride ions, and it is here that the ion-exchange screens play an important role in suppression capacity. They provide a high concentration of available ion-exchange sites in the hydroxide form. This enhances the reaction that forms water and creates an electrical field.

For anion analysis, the suppressor has cation exchange membranes and screens. In the simplest case, NaOH is the eluent. The two suppression mechanisms are analogous to the cation analysis. With anions, the $Na^+$ in the eluent exchanges with $H^+$ on the membrane in the first mechanism. The second mechanism is based on the fact that water is a much weaker acid than the ion-exchange site on the anion micromembrane suppressor. Water is formed from eluent hydroxide and a H from the membrane-ion exchange site. This causes an electrical field where there is excess positive charge in the eluent (from $Na^+$) and excess negative charge in the membrane (unfilled anionic exchange sites). The sodium ions move under the force of this field. They fill the ion exchange sites on the membrane and neutralize the charge imbalance.

Quite analogous to the cation membrane suppressor, the anion membrane suppressor probably utilizes both mechanisms for suppression. At the suppressor inlet, there is a greater chance for $Na^+$ in the eluent to exchange with $H^+$ in the membrane. Near the exit of the suppressor, the formation of water becomes significant. It is here that the high concentration of available ion exchange sites in the hydrogen form play an important role in suppression.

This information can be useful when considering eluents to be used. For example, compounds such as amino acids that form zwitterions can be used as eluents. If its isoelectric point is between pH 5 and 8, the amino acid would have a negative charge at normal eluent pH (above 9). Thus, it can function as an eluent for anions. As it enters the suppressor, several things can happen. The acidic environment of the suppressor can protonate the molecule, making it a cation. As a cation, it can pass through the membrane via cation exchange sites. Alternatively, the molecule can be protonated back to its isoelectric point. The neutral molecule can either diffuse through the membrane into the regenerant stream or exit the suppressor as a neutral molecule with low conductivity. The degree to which this happens is related to the isoelectric point of the zwitterion, the rate at which it transports through the membrane, its concentration in the eluent, and the pH of the suppressor effluent. For a zwitterion to be adequately suppressed, 50 to 75% must be removed by diffusion through the membrane of the suppressor. The background conductivity in a zwitterion-containing eluent is dependent on the percent removed by the suppressor, and to how close its isoelectric point matches the suppressor pH. All the amino acids with an isoelectric point (pI) between 5 and 8 represent potential eluents. The closer the pI is to 7, the better the zwitterionic amino acid will be suppressed. One such amino acid, tyrosine, is not stable to ambient light and temperature for more than a few days. Glycine is quite stable and can be a good eluent for cations and anions.

When a micromembrane is operating properly and is at equilibrium, there are three definable regions longitudinally in the membranes of the suppressor and eluent screen. At the inlet, the eluent screen and membranes are completely in the sodium form due to the initially overwhelming concentration of sodium in the eluent. As the eluent flows through the eluent chamber, there is a gradient from high to low sodium concentration and a proportional reverse concentration gradient of hydronium ions associated with the ion exchange sites. If the suppressor is adequately suppressing the eluent, there is a region of screen and adjacent membrane that is completely in the hydronium ion form. Finally, as a band passes through the suppressor, cations associated with the strong acid analyte anions are exchanged for hydronium ions, significantly increasing the total conductance of the sample band by a factor of 5 to 7, further enhancing sensitivity.

When a weak eluent is used, the first region, where the membrane is in the sodium form, is relatively small. The totally regenerated region, however, is quite large. When a strong eluent is used, the region of the membrane that is in the sodium form is considerably larger, and the totally regenerated region is small. This is true when two different eluent concentrations are used, and also when a gradient elution is used. The effect of varying NaOH eluent concentration during a gradient elution is shown in Figure 3 for an anion micromembrane suppressor which has a dynamic suppression capacity of 200 meq/min. The initial concentration of 50 m$N$ NaOH flowing at 1 m$\ell$/min uses up only a fraction of the cation exchange sites on the membrane suppressor. As the gradient increases to 100 m$N$ NaOH, more cation exchange sites are used up until at 150 m$N$ NaOH, two-thirds of the sites are used up. Because the suppressor has a high capacity, it responds quickly to changes in eluent

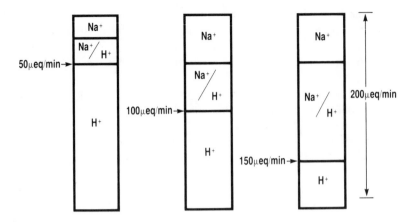

FIGURE 3.    Effect of eluant concentration on suppressor ion exchange form. Illustrated for anion suppressor.

concentration, providing that the gradient remains within the suppressible concentration range.

To adequately suppress the conductivity of the eluent, a proper regenerant must be chosen. Just as dynamic suppression capacity can be calculated, so can dynamic regenerant concentration. It is defined as the product of regenerant flow-rate in milliliters per minute and regenerant concentration in microequivalents per milliliter. As a general rule, the dynamic regenerant concentration should be three to five times the dynamic suppression capacity. Three major characteristics of the regenerant affect the suppressor capacity:

1.    The nature of the regenerant
2.    Regenerant concentration
3.    Regenerant flow rate

The regenerant should be a strong acid (pKa < 1) because it must be a source of hydrogen ions. The acid co-anion should be large and/or polyvalent so that it is effectively excluded from the membrane causing minimum co-leakage into the eluent stream. This is why sulfuric acid is such a popular regenerant. Octane sulfonic acid could also be used as a regenerant. It has a larger co-anion than sulfate, but is a much weaker acid than sulfuric acid. This means that a higher concentration of octane sulfonic acid would be required for the same suppression capacity as sulfuric acid.

Unfortunately, regenerant concentration is limited by the membrane Donnan potential. Above a critical regenerant concentration, this potential is overcome and co-anions are no longer excluded from the membrane. Because dynamic regenerant concentration is also controlled by regenerant flow rate, this problem can be partly overcome by keeping the regenerant concentration below the critical limit and increasing the flow rate. In fact, less noise is seen in the baseline if the regenerant concentration is low, but the flow rate high.

One of the major advantages of the micromembrane suppressors is that they enable the use of eluents that could not be adequately suppressed with packed bed or fiber suppressors. For example, if one wanted to separate fluoride, acetate, formate, and chloride on an AS-1, 2, 3, or 4, an eluent weaker than the standard bicarbonate/carbonate is required. Sodium tetraborate at concentrations from 5 to 10 m$M$ would be adequate.[32] If the sample also contained nitrate and sulfate, much higher borate concentrations would be required. Conventional packed-bed and fiber suppressors are not capable of suppressing the high borate concentrations that would be needed. Another eluent that is weak enough to separate fluoride,

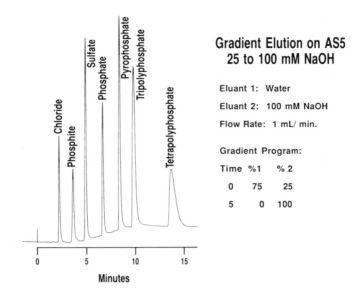

FIGURE 4.   Gradient elution on AS5. Gradient is 25 to 100 m*M* NaOH.

formate, acetate, and chloride is NaOH (at about 5 to 10 m*N*). As with borate, however, much higher concentrations of NaOH would be required to elute nitrate, sulfate, or other polyvalent anions. This can be accomplished by using a gradient elution with an anion micromembrane suppressor. As illustrated in Figure 4, mono-, di-, tri-, and tetravalent anions can be separated using a gradient of 25 to 100 m*M* NaOH.

When choosing an eluent, both chromatography and suppression should be considered. From a chromatographic point of view, it should be remembered that like elutes like. Monovalent anions are better eluted with a monovalent eluent. Organic anions more effectively elute organic anions. One should consider eluent pH and suppressor pH, especially when using zwitterionic eluents. In addition, eluent-column compatibility must be considered. Not all ion chromatography separator columns are resistant to organic solvents at high concentrations. Polystyrene that is only lightly cross-linked tends to shrink and swell as the amount of organic solvent in the eluent changes. This causes the resin bed to be unstable, and unwanted channels or voids can develop. For example, tetrahydrofuran, a common solvent for gel permeation chromatography, is unacceptable to most ion chromatography columns.

The micromembrane suppressor provides several practical advantages. Because of the extremely small void volume and high suppression capacity, a more stable baseline is obtained at sensitive (0.01 μs full scale) detector settings. This lowers the detection limits for common anions as illustrated by the analysis, without prior sample preconcentration. The higher dynamic suppression capacity permits the use of stronger, more concentrated eluents such as 20 m*M* NaOH and 3 m*M* glycine, which could not be used with packed bed or fiber suppressors. This also shortens analysis time and increases sensitivity because the peaks are sharper. This is especially true for strongly retained ions. Another advantage of the micromembrane suppressor is its stability to organic solvents and detergents used as ion pairing reagents in mobile-phase ion chromatography (MPIC). Thus, anions such as metal cyanide complexes can be determined by MPIC without the need for packed bed suppressors described in the literature.[32,33]

The micromembrane suppressor (MMS) also enabled the development of gradient ion chromatography. The limited capacity of the fiber suppressors prevented them from adequately suppressing the increasing eluent concentrations used in a gradient. The increased capacity of the MMS is able to adequately suppress the increased eluent concentration.

Thus, the MMS illustrates a point that needs repeating. Almost none of the papers published before 1986 use a MMS, or even many of the newer separator columns. This in no way lessens the validity of the data in these reports. The MMS and other advances in ion chromatography hardware provide faster, more sensitive analyses and, in some cases, provides enhanced analytical capabilities. Undoubtedly, laboratories will continue to publish reports using fiber suppressors, isocratic elution, and older separator columns. These systems still perform quite well and are fully capable of producing valid analytical data.

## IV. ANION DETECTION WITHOUT CHEMICALLY SUPPRESSED CONDUCTIVITY

In the midst of all the developmental work that was being conducted on improved suppressor columns, other detection methods were developed. A conductivity detector is useful for anions only if the anions have sufficiently high specific conductance at the pH of the regenerant effluent. This means that the anions should have a pKa less than 7. If their pKa is above 7 (i.e., cyanide), they will not be detected by chemically suppressed conductivity. For example, when detecting chloride by chemically suppressed conductivity, both of the ions, $H^+$ and $Cl^-$, are actually being seen simultaneously by the conductivity detector. The hydrogen ion has the highest specific conductance of any cation, so the conductance of the analyte anion, chloride, is enhanced (because it is detected as HCl and not as a salt). Cyanide, on the other hand, is converted to HCN by the suppressor. Since HCN is not appreciably dissociated, it has very low conductivity and cannot be detected by the conductivity cell. Cyanide will react with a silver electrode to produce a measurable current. This type of electrochemical detection is quite sensitive, so that parts per billion levels of cyanide can be detected without prior sample preconcentration. Similarly, sulfide, bromide, sulfite, iodide, thiocyanate, and persulfate can also be detected using a silver electrode. Alcohols, glycols, hydrazine, and hypochlorite can be detected using a platinum electrode. Phenols, catechols, and nitrobenzenes can be detected using a glassy carbon electrode. Sugar alcohols, along with mono-, di-, and oligosaccharides, can be detected using a pulsed amperometric detector (see Chapter 3 for details).

For many anions which cannot be detected easily by electrochemical or conductivity detectors, post-column reaction methods were developed. The first post-column reactor for anions was based on the color formed when ferric nitrate reacts with polyvalent anions such as polyphosphates, phosphonates, and chelating agents such as EDTA and diethylenetetramine penta acetic acid (DTPA). The AS-7 column with a 0.03 to 0.07 $N$ nitric acid eluent (concentration depending on the anions being determined) and an UV-visible detector was used to separate such anions.[34] Since then, a method based on anion exchange was developed to separate amino acids on one column in one run. If the secondary amine, proline, is also to be determined, the post-column reaction is with ninhydrin. If only primary amines are being determined, the post-column reaction is with *o*-phthalaldehyde. The variety of different detection schemes available will be discussed more in Section VIII and in Chapter 3.

It is important to realize, though, that ion chromatography has become a technique where the analyst can mix and match appropriate separation and detection methods to optimize the analysis. Because detection methods other than chemically suppressed conductivity are relatively new, their importance should not be understated. However, there are many very important anions and cations whose only physical property is one that can be sensitively detected. Examples include the common anions fluoride, chloride, phosphate, and sulfate, and cations such as ammonium. For these ions, chemically suppressed conductivity is the detection method of choice.

## V. ADVANCES IN CATION ANALYSIS

One of the first advances was to discover an eluent for separating divalent cations, calcium, magnesium, barium, and strontium.[35,36] The eluent was 2.5 m$M$ $m$-phenylenediamine · 2 HCl plus 2.5 m$M$ nitric acid. The CS-1 separator column was used. The eluent is considerably stronger than the 5 m$M$ HCl used to separate monovalent cations. Thus, lithium, sodium, ammonium, potassium, and cesium all coelute in the void volume when the $m$-phenylene-diamine eluent is used. In fact, the $m$-phenylenediamine is retained so strongly that it is never completely washed off the CS-1 with 5 m$M$ HCl. Thus, a CS-1 column had to be dedicated to either mono- or divalent metal determinations. It was impossible to simultaneously separate mono- and divalent cations.

The next major advance in cation analysis was the development of the CS-2 column for separation and post-column reaction for separation of transition metals and a post-column reactor for delivery of 4-pyridyl azo resorcinol (PAR), which forms highly colored metal-PAR complexes.[37] The post-column reactor utilizes a membrane reactor where effluent off the separator column enters at the top, and PAR enters at the bottom on the outside of the membrane. The PAR diffuses across the membrane to bind to the transition metals. This effluent is then directed to a UV-visible detector which is set at 420 nm.

The post-column reaction was needed for detecting transition metals because of a problem with using chemically suppressed conductivity. The suppressor columns all convert metals to their hydroxides. The transition metal hydroxides are insoluble, and never reach the conductivity cell.

Elution of the transition metals off the CS-2 is based on a slightly different principle than that used to elute cations off the CS-1 or inorganic anions off the various anion separators. All the transition metals have almost identical affinities for cation exchange sites. If one uses an eluent such as HCl, which merely competes for ion exchange sites, the transition metals will not be separated. On the other hand, transition metals differ from one another in their ability to bind mild chelating agents,[38] i.e., their dissociation constants are different. Thus, by using mild chelating agents such as oxalic and citric acid in the eluent, the transition metals can be separated. The mechanism is not competition for ion exchange sites, but is based on differential binding to chelating agents in the eluent.

The next major advance in cation analysis was the development of a specialty cation exchange resin, the CS-4, for simultaneous separation of amino acids in one run. The amino acids are detected by post-column reaction with either ninhydrin or $o$-phthalaldehyde as described previously for the anion exchange method. The amino acids were separated as cations using a four-step eluent gradient at low pH (see Chapter 3 for details).

The next cation separator column introduced was the CS-5 for improved transition metal separation.[36] The CS-5 column uses a pyridine-2-carboxylic acid eluent in a nonoxidizing environment for accurate speciation of ferrous and ferric ions. The CS-5 is also more efficient than the CS-2, producing sharper peaks and lower detection limits. It has been used in the simultaneous separation of lanthanide and transition metals.

The next cation separator introduced was the CS-3, which was designed for simultaneous separation of mono- and divalent cations in one run. Sodium, ammonium, and potassium are more strongly retained on the CS-3 than the CS-1 using 5 m$M$ eluent. A stronger eluent is required to elute them in the same time that they are eluted off the CS-1. Thus, stronger eluents are used with the CS-3. The mono- and divalents can be separated using a single step gradient. Alternatively, they can be separated using a linear gradient (see Chapter 2 for details). The cation micromembrane suppressor (CMMS) makes it possible to use the stronger eluents and gradient elution itself. It is useful to expand the capacity for cation analysis. Organic cations such as aliphatic and aromatic amines can be separated either by ion exchange or by ion pair chromatography. In the case of aliphatic amines, which have little or no UV absorbance, chemically suppressed conductivity with the CMMS can offer a distinct advantage over other analytical methods.

Again, few publications and conference papers on the ion chromatography of cations have described the use of the newest separators. The results obtained in such studies are certainly valid and laboratories already equipped with the older columns will continue using them to produce valid analytical data. Thus, it is important to understand the basic principles of ion chromatography to be able to critically evaluate reported results on similar samples using different columns.

Although the bulk of the present literature on ion chromatography with chemically suppressed conductivity detection deals exclusively with anion analysis, the capabilities for inorganic and organic cation analysis are noteworthy. The monovalent and the alkaline earth metal cations can now be separated and detected simultaneously in one analysis. Organic cations, especially the aliphatic mono-, di-, and even polyamines, can also be separated by cation separators and detected by chemically suppressed conductivity (and other methods that will be discussed later).

## VI. MOBILE-PHASE ION CHROMATOGRAPHY

The separator columns described so far are usually used exclusively for either cations or for anions. There is a column, however which can be used for both anions and for cation separations. This column is the mobile-phase ion chromatography (MPIC) column. The term "mobile-phase ion chromatography" is also used to describe the overall analytical technique. This technique uses a separator column which contains no permanent ion exchange sites. It is a nonionic column with polarity intermediate between silica and C-18 (octadecyl)-silica columns commonly used in HPLC.

In fact, the technique of MPIC does closely resemble the well-known HPLC technique of reversed-phase ion pair chromatography. It is important, though, to understand the similarities and differences between MPIC and reversed-phase HPLC. To begin with, MPIC is a trademark designating that the macroporous polystyrene/divinylbenzene-containing separator column was manufactured by Dionex. Columns used in reversed-phase HPLC are typically silica based with a long-chain (C-12 or C-18) alkyl group chemically bonded to the silica. However, there are some reports in the literature about other macroporous polystyrene/divinylbenzene resins for reversed-phase HPLC.[39-42] Thus, one rather general definition of reversed-phase HPLC is "any form of liquid chromatography in which the mobile phase is more polar than the stationary phase". This definition does not specify whether the stationary phase is C-18 silica or polystyrene/divinylbenzene. It does not specify whether the analytes are ionic or nonionic. It does not specify the detection method being used. It could be electrochemical, post-column reaction with UV or fluorescence, conductivity, or anything else. Using this definition, MPIC would also be a type of reversed-phase HPLC. The MPIC column is nonionic; it has no ion exchange sites. Although it was developed originally for use with chemically suppressed conductivity detection to determine anions, the method uses the well-known HPLC technique of ion pair chromatography.[43]

On the other hand, the terms MPIC and reversed-phase HPLC are usually used in very different contexts, and, to most scientists, denote different technologies. MPIC is almost always described in reports that include descriptions of other ion chromatography methods using anion separators and suppressors.[44] Thus, MPIC is usually considered to imply the use of chemically suppressed conductivity and the chromatographic separation of ions. Most reversed-phase HPLC separations, on the other hand, usually use acetonitrile/water or methanol/water eluents, silica-based columns, and UV detection. In many cases, the compounds being determined are nonionic. Although silica-based columns (and non-MPIC polystyrene/divinylbenzene columns) have been used in the chromatography of ions,[45-47] they have not used chemically suppressed conductivity detection. Moreover, when one mentions reversed-phase HPLC to most scientists, they think of the determination of polar organic compounds; when one mentions MPIC they think of the chromatography of ions.

Thus, when discussed in its most common context, MPIC almost suggests that ions are being separated and suppressor columns are being used for detection. MPIC columns would have a rather limited applicability if that was all they were capable of. In fact, MPIC columns can also be used for ion pair chromatography of cations, especially aliphatic amines, which are difficult to separate and detect by other methods. It is surprising that more methods have not been published on MPIC determinations of amines. Perhaps its reputation for anion analysis has kept MPIC from being carefully investigated for cation analysis.

Just as one is not restricted to anion separations by MPIC, it is foolish to feel restricted to the use of conductivity detection. MPIC with electrochemical detection has been shown to be quite capable of determining phenols and nitrobenzenes.[48] Thus, MPIC can be useful in separating nonionic compounds and the most useful detector may well be the UV-visible detector that is so popular in reversed-phase HPLC.

Reports are just now beginning to appear which show the use of MPIC with UV-visible detection for the determination of nonionic compounds.[49,50] The pH stability of the poly-styrene/divinylbenzene is quite useful when analyzing highly acidic or alkaline solutions used in the electronics and electroplating industries. Typically, organic additives are present at low concentrations in such solutions, which means that samples as corrosive as 2% NaOH or 2% $HBF_4$ have to be injected (see Chapter 4). Such extremes of pH would be harmful to silica-based columns.

Thus, when reading the existing literature on MPIC, the principles of ion-pair chromatography and chemically suppressed conductivity should be kept in mind. However, when using MPIC in the laboratory for future work, it should be remembered that MPIC can be used effectively to separate nonionic organics as well.

## VII. OTHER DETECTION METHODS

Not only are there other separation techniques, but also other methods of detection in ion chromatography. The original invention implied the use of ion exchange separator columns and chemically suppressed conductivity detection. In fact, separations need not be based on ion exchange, and can utilize MPIC. Similarly, there are some ions which are not conveniently detected by conductivity, but can be detected by other means. For example, anions with pKa above 7 are only very weakly ionized after coming off the suppressor. These un-ionized anions, such as cyanide and sulfide, exist as HCN and $H_2S$, which have very low specific conductances. Both sulfide and cyanide will react at a silver electrode to produce AgCN and $Ag_2S$, and an easily measurable current. This is the basis for the earliest electrochemical detection methods. A compound or ion will undergo an oxidation at an electrode surface, producing a current that is proportional to its concentration. Other anions which will react at a silver electrode include bromide, iodide, thiosulfate, and thiocyanate. In each analysis, a fixed voltage is applied between the silver working electrode and a Ag/AgCl reference electrode. The method is very sensitive. For example, cyanide can be detected at parts per billion levels.

Even before their use in ion chromatography, electrochemical detectors were well known in liquid chromatography. They typically operated by applying a single fixed potential between a working electrode and a reference electrode in a flow-through cell. They used a three-electrode cell, consisting not only of the working and reference electrodes, but also a counter electrode. For example, a glassy carbon counter electrode allows the potentiostat to keep the applied potential constant, and prevents any damaging current drain that could occur at the reference electrode. The electrochemical detector is placed after the separator column. As each electroactive analyte comes off the column and passes through the detector, a current is measured. Within some limits, this current is directly proportional to the concentration of electroactive analyte. In developing a new method, the applied voltage is adjusted to maximize the signal to noise ratio, and hence the sensitivity.

Various other electrodes can be used for detecting ions and compounds not conveniently detected by a silver electrode. Phenols, aromatic amines, and nitrobenzenes will react at a glassy carbon electrode when the proper voltage is applied. Hydrazine, hypochlorite, and alcohols will react at a platinum electrode. In addition, a Ag electrode is sensitive to changes in pH and can be used for indirect detection of a number of ions that are not usually considered electroactive. These anions are not oxidizable, but they do cause a pH change when they elute off the suppressor column. It should be remembered that the eluent is converted to $H_2CO_3$, which has a pH of about 4. When fluoride, chloride, phosphate, nitrate, and sulfate elute, they also come off in their acid forms. HF, HCl, $H_3PO_4$, $HNO_3$, and $H_2SO_4$ all are strong acids. As a result, they are almost completely dissociated and cause a drop in pH when they elute off a separator column. Thus, by setting the electrochemical detector just past an anion fiber suppressor, Tarter[51] was able to detect these anions. It should be noted that, for the usual determinations of electroactive species, a suppressor, column is not used. For example, the eluent for cyanide and sulfide is 1 m$M$ sodium carbonate, 10 m$M$ $NaH_2BO_3$, and 15 m$M$ ethylenediamine. This eluent also provides the supproting electrolytes needed to effect the electrochemical reaction. To obtain optimum signal response, it is important that this eluent not be put through an anion suppressor column. The carbonate and borate would be converted to carbonic acid and boric acid. The ethylenediamine would be completely removed from the eluent (especially if a high-capacity packed bed or micromembrane suppressor were used). Thus, it is important to remember that putting the electrochemical cell in front of or past a suppressor column will dramatically affect its response to analytes. As a rule of thumb, the electrochemical detector is placed in front of the suppressor column, or, alternatively, it is possible to not even use a suppressor column.

In the examples cited for electrochemical detection with a single fixed applied potential, the sample contacts the electrode surface. This causes no problem when the reaction products do not deposit on the electrode surface and change its characteristics. Some electroactive species, most notably carbohydrates, do form oxidation products which poison the working electrode. This causes excessive noise, decreased signal, and baseline drift. Thus, it was necessary to develop a means to clean the electrode surface. One alternative, of course, would be to disassemble the electrochemical cell after each chromatogram and polish the electrode surface. A much better alternative would be to clean the electrode surface by automatically applying an appropriately high voltage to strip the oxidation products off the electrode surface. In this way, repetitive analyses are made possible with a minimum of applied labor being required.

An electrochemical detector was invented with just this purpose in mind: to be able to apply a series of voltage pulses to first generate a current, then to clean the reaction products off the electrode. Invented by Hughes and Johnson, the pulsed amperometric detector can now be used to detect subparts per million levels of carbohydrates and polysaccharides.[52] These sugar alcohols will react at a gold electrode when a potential of about 0.2 mV is applied. The oxidation products are removed by using a pulse of 0.5 mV followed by a $-0.2$-mV pulse. In this way, the gold electrode surface is cleaned so that it will respond reproducibly to the next carbohydrate sample.

Electrochemical and chemically suppressed conductivity detectors are not the only choices available. The analyst is also able to choose UV-visible and fluorescence detectors. In ion chromatography, these detectors are usually placed just past a post-column membrane reactor. Ions such as amino acids, transition metals, and polyvalent anions will react with a specific spectrophotometric or fluorometric reagent to produce either a colored or a fluorescent product. This product can then be detected based on either its UV or visible absorbance, or its emission of fluorescent light. The first application for this was the detection of transition metals. Transition metals, like all other cations, are converted to their hydroxide form by the cation suppressor. The transition and lanthanide metal hydroxides are insoluble, and never reach the conductivity cell. Thus, they cannot be detected based on chemically suppressed conductivity. Instead, the analyst takes advantage or the fact that 2-pyridyl-4-azo resorcinol (PAR) will react rapidly under proper conditions with transition and lanthanide metals to form a highly colored complex. Some of these complexes have slightly different wavelengths for maximum absorbance, but they do all absorb very well near 520 nm. Thus, the UV-visible detector is set at 520 nm, and becomes an almost universal detector for metal-PAR-colored complexes.

The distinct advantage of this method is partly due to the membrane reactor. It is analogous to the fiber suppressors discussed earlier, except that it is reagent and eluent that flow in opposite directions on opposite sides of the membrane (as opposed to reagent and regenerant in a fiber suppressor). The PAR reagent permeates through the membrane so that it can react with the metal ions as they elute off the cation separator column. A colored PAR-metal complex forms rapidly, and flows to the UV-visible detector which has a flow-through cell. By using a membrane to mix the spectrophotometric reagent with the analyte, there is no increase in elution volume so typical of conventional T-shaped mixing cells. The concomitant peak broadening is avoided by using the membrane reactor. Thus, the post-column membrane reactor enables reagent and analyte to mix with little or no loss of chromatographic efficiency.

This same post-column membrane reactor was then used to detect polyvalent anions which were strongly retained on most anion separator columns. Polyphosphates, polyvalent chelating agents, and polyphosphonates required concentrated eluents for their elution. The AS-1, 2, 3, 4, and 5 separator columns all had limited capacity, as did the fiber suppressors. As a result, the strong eluents needed for polyvalents would tend to overload the separator, and not be sufficiently suppressed by the fiber suppressor. Thus, a new column with a higher ion exchange capacity, the AS-6, was developed. This column was not overloaded with the 0.15 $N$ HNO eluent required for polyvalent anions. In addition, in the landmark paper by Inamari et al.,[53] who described the spectrophotometric detection of over 25 anions by their reaction with ferric nitrate. The Fe forms a colored complex with these anions. Fortunately, the polyvalent anions all form this complex rapidly and all absorb strongly at 350 nm. As a result, the post-column reagent for the ion chromatographic method consists of 1 g/$\ell$ $Fe(NO_3)_3 \cdot 9H_2O$, and the eluent can vary from 0.05 to 0.15 $N$ $HNO_3$. The polyvalent anions can be detected at parts per million levels using this approach.

Probably the most important and one of the oldest post-column reaction methods for liquid chromatography is the use of ninhydrin to detect amino acids.[54] Another very useful post-column reagent for amino acids is $o$-phthalaldehyde, which reacts with primary amines to form a fluorescent product.[55] Certainly this post-column reaction can be used in the ion chromatographic determination of amino acids. The ninhydrin is used when determining primary and secondary amines. Thus, if determining proline, a ninhydrin and a UV-visible detector are used. If determining only those amino acids which are primarily amine, $o$-phthalaldehyde and a fluorescence detector are used. The amino acids can be separated as cations (on a CS-4 column) or as anions (on an AS-8 column) simultaneously in one run (see Chapter 2 for details). Only one column is needed. The choice of whether to separate the amino acids as cations at low pH or as anions at high pH, depends on the application and the sample matrix. In either event, the post-column chemistry is well known. The same ninhydrin or $o$-phthaldehyde that have been used by biochemists for years are used. Because the reaction product with ninhydrin absorbs in the visible, a UV-visible detector is used. Ninhydrin reacts with primary and secondary amines, and with ammonium. The fluorescence reagent *o*-phthalaldehyde is used for primary amines. The ninhydrin has an advantage of being a more universal reagent for amino acids, but *o*-phthalaldehyde is more sensitive since fluorescence is inherently more sensitive than UV-visible absorbance. The choice of post-column reagent depends on whether proline is to be determined and on the sensitivity needed for the other amino acids.

The amino acid analysis by ion chromatography is especially useful because it can be automated. Whether separated as anions or cations, a series of step gradient elutions is required. This switching of eluents can be done automatically in a fully unattended mode. All the analyst need do is prepare the sample, load it into an autosampler, and recall previously written programs on a computer or microprocessor. The automated ion chromatograph will select the proper chromatographic conditions, make necessary calculations, and print out results in report format.

In fact, almost any ion chromatographic analysis can be automated. It is important to understand the hardware of the ion chromatograph to ensure that all instrument parameters are automated; but, as with any automation, it is most applicable when analyzing many identical samples frequently. Because of the extreme versatility of an ion chromatograph, it is also possible to use automation to perform several types of analyses on different samples requiring very different data. It is important to understand the basic principles of ion chromatography, so that compatible analyses are performed in an automated analysis. For example, separations of small, hydrophilic inorganic anions often uses a bicarbonate/carbonate eluent. Separations of aliphatic carboxylic acids by ion chromatography exclusion (ICE) utilize acidic eluents (i.e., 1 m$M$ HCl). If the carbonate eluent was to mix with the HCl eluent, carbon dioxide gas bubbles would evolve, harming chromatographic performance. On the other hand, monovalent cations use acidic eluents (i.e., 5 m$M$ HCl on the CS-1) and their chromatography would be compatible with ICE separations that also use HCl eluent. If the analyst already knows the eluents that are needed for each analysis, the ion chromatograph can be plumbed in such a way that only compatible eluents are used in a single automated analysis. By careful planning of work loads, the analyst can predict which analyses are required so that the proper columns and detectors can be installed in their proper locations on the ion chromatograph. To do this efficiently, it is necessary to understand the components of the ion chromatograph that are controlled by the microprocessor or computer.

An ion chromatograph can consist of one or two identical systems. Each system is equipped with six different eluent ports connected to an analytical pump (Series 2000) or a gradient pump (Series 4000). When doing isocratic or step gradient elution, the proper eluent can be selected. If doing step gradients, the microprocessor or computer can automatically switch eluents at the prescribed times. Alternatively, if performing two different analyses requiring two different eluents, the proper eluent can be switched once the first analysis is finished. The column can be equilibrated with the second eluent, then the first sample injected after a preset equilibration time. The pump speed can also be automatically adjusted so that the desired eluent flow rate is obtained. There is also a load/inject valve and an autosampler that are under computer control. The position of the autosampler is controlled so that the desired standard or sample is selected. The computer sends a signal to a small pipet arm that dips into the desired sample. A pump on the autosampler is then switched on, and the load/inject valve is placed in the load position. After pumping sample solution for about 1 min, the load/inject valve is well rinsed and loaded. The computer then switches the load/inject valve to the inject position, injecting the sample. The ion chromatograph also has a chromatography module that houses the separator and suppressor columns. One particularly useful arrangement is to have two columns connected to one valve, the A valve. This valve is positioned just downstream from the load/inject valve. The microprocessor or computer can select the desired separator column by switching the A valve off or on. For example, as AS-4A column might be installed so that it is selected when the A valve is on. The AS-

4A can be used to separate small, hydrophilic anions using a bicarbonate/carbonate eluent. An AS-5 column can be installed so that it is selected when the A valve is off. It can be used to separate large, hydrophobic anions using the same bicarbonate/carbonate eluent used on the AS-4A column. All that is necessary when switching from AS-4A analyses to AS-5 analyses is to have the software written so that the A valve goes from on to off. The eluent is not changed. It is simply redirected to the AS-5. After about 15 min of equilibration, the AS-5 is ready to be standardized. The effluent line coming off the AS-4A and the AS-5 are both directed to the same anion suppressor column through the B valve. This B valve is also used to control regenerant flow. If the B valve is on, regenerant flow is enabled, and if the B valve is off, regenerant flow is blocked.

A computing integrator or a computer itself, can be easily programed so that it knows all the pertinent information about standards and samples. The computer is told the number of ions that are present in each standard along with their concentrations. The computer is told to fit the number of standards to be injected. After the prescribed number of injections, the computer can then automatically fit the calibration data to a straight line (or perform nonlinear curve fitting to higher degree polynomials if desired). Because the program has been written so that the computer "knows" when the last standard has been injected, it makes the necessary calculations and is ready to analyze standards. The analyst enters the name of each standard and its dilution factor or sample weight as part of the initial method setup. This, along with all other steps, are entered into the computer's memory through a menu-driven program. Little or no programing experience is needed. The computer can then use the calibration data, together with the chromatographic data on each sample to make further calculations. The identity of each ion in the sample is confirmed if its elution-time is close enough to that of the standard ion. The definition of "close enough" is decided by the analyst when writing the method software. The concentration of any ion identified by its retention-time is then calculated based on calibration data, dilution factors, and/or sample weight. The results are printed out in report format. A more detailed description of how to carefully plan analyses so that compatible eluents are used, along with specific applications for automation, is given in Chapter 3.

## VIII. BASIC DEFINITIONS IN ION CHROMATOGRAPHY

Ion chromatography is simply a form of chromatography, so the same basic principles and definitions that are used in gas chromatography and HPLC apply.[56] Consider a sample containing three components. Prior to injecting the sample, the separator column is equilibrated with solvent (i.e., eluent). The separation process can be illustrated in general terms as seen Figure 5. The dotted areas represent the original eluent in the column. As the sample reaches the column, it begins to displace eluent off the column. When doing ion exchange, this "displacement" amounts to ion exchange. As the sample flows down the column, it begins to separate into its individual components. The separation is based on the fact that each component has a different affinity for the separator column.

This affinity can be described as a distribution coefficient, K. It is simply the concentration of a given component in the stationary phase, $[X]s$, divided by its concentration in the mobile phase, $[X]m$. The length of time that a component stays on the separator column is referred to as its retention, or capacity factor, $k'$. It is defined as the number of column volumes necessary to elute the given component. In other words, every column has a certain void volume. For example, there are always a few voids between the sulfonated polystyrene/divinylbenzene in a cation-separator column. The retention of a cation on this separator would be directly related to its retention time, $t_r$, or its retention volume, $V_r$. For component number 1 in Figure 6, let us define its retention volume as $V_1$, and the column void volume as $V_o$. The capacity factor for component 1 would be $k' = (V_1 - V_o)/V_o$. It is simply the

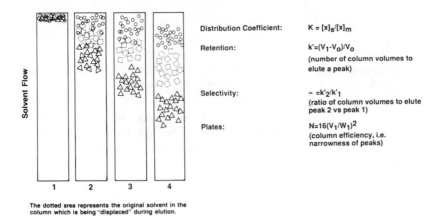

Distribution Coefficient:  $K = [x]_s/[x]_m$

Retention:  $k' = (V_1 - V_0)/V_0$
(number of column volumes to elute a peak)

Selectivity:  $\propto = k'_2/k'_1$
(ratio of column volumes to elute peak 2 vs peak 1)

Plates:  $N = 16(V_1/W_1)^2$
(column efficiency, i.e. narrowness of peaks)

Solvent Flow

1  2  3  4

The dotted area represents the original solvent in the column which is being "displaced" during elution.

FIGURE 5.   Hypothetical separation of a three component mixture. Circles, squares, and triangles represent three different components.

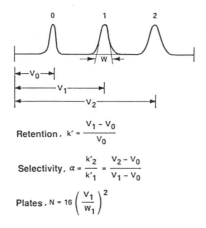

Retention, $k' = \dfrac{V_1 - V_0}{V_0}$

Selectivity, $\alpha = \dfrac{k'_2}{k'_1} = \dfrac{V_2 - V_0}{V_1 - V_0}$

Plates, $N = 16 \left(\dfrac{V_1}{W_1}\right)^2$

FIGURE 6.   Retention-selectivity-plates. Equations are given for retention (also called capacity factor), selectivity, and plates (a measure of column efficiency).

number of column volumes needed to elute a peak. The primary objective in chromatography, though, is not to just elute a component, but to separate it from other components. The efficiency of a column can be measured in terms of its ability to resolve the individual components. The term ''selectivity'' is used to describe the ratio of column volumes needed to elute peak 2 vs. peak 1. The column selectivity is directly related to its efficiency.

Efficiency was first measured before the advent of column chromatography. At that time, distillation was a common means of separating components in a mixture. It was discovered that filling the condenser tube (in the distillation apparatus) with glass plates increases the efficiency of the separation. Thus, the efficiency of the separation was found to be a measure of the number of glass plates in the condenser. Because the plates could vary in surface area, it became necessary to define a ''theoretical'' plate which would be standard for all laboratories. This concept of a standard theoretical plate was so useful that it has survived to this day when describing modern chromatographic column efficiency. The number of theoretical plates, N, is given by $N = (V_1/W_1)$, where $W_1$ is the width of the peak produced by component 1. Ideally, the number of theoretical plates in a column should not vary significantly from one component to another. In practice, one anion, such as chlorobenzoic acid, may have different interaction with an anion separator than sulfate. The chlorobenzoic acid peak may exhibit peak tailing due to nonionic interactions with the polystyrene/divinylbenzene, whereas the sulfate does not. As a result, there may not seem to be as many theoretical plates for the chlorobenzoic acid as for the sulfate. As a result, when mentioning the number of theoretical plates in any publication, the peak used to make the measurement, should be listed.

Many articles on ion chromatography and HPLC use the term "capacity factor". It is a measure of the number of column volumes needed to elute a peak. It can be calculated from either the retention time, $t_r$, or the retention volume, $V_r$, as shown below:

$$k' = \frac{V_r - V_o}{V_o} = \frac{t_r - t_o}{t_o} \tag{1}$$

The capacity factor is simply the mass of the analyte in the stationary phase, $M_s$, divided by the mass of the void volume, $M_o$. This in turn is related to the concentration in the stationary phase, $C_s$, and concentration in the mobile phase, $C_m$:

$$= \frac{M_s}{M_o} = \frac{c_s \cdot v_s}{c_m \cdot v_o} \tag{2}$$

The ratio of concentrations in the stationary phase and mobile phase is simply the distribution coefficient, K. Thus, the capacity factor is related to the distribution coefficient by the equation:

$$= K \cdot \frac{v_s}{v_o} \tag{3}$$

The column selectivity can also be related to the distribution coefficient. Selectivity is simply a measure of how well peak maxima are separated. It can be expressed as:

$$\alpha = \frac{t_2 - t_o}{t_1 - t_o} = \frac{k'_2}{k'_1} = \frac{K_2}{K_1} \tag{4}$$

In other words, selectivity for two different analytes is simply the ratio of their capacity factors, $k'$, or their distribution coefficients, K.

The column efficiency, which is a measure of the sharpness of the bands, is expressed in terms of theoretical plates, N:

$$N = 16 \left(\frac{t_r}{w_b}\right)^2 \tag{5}$$

where $t_r$ is the retention time and $W_b$ is the width of the peak at the baseline. These are measured form a sample chromatogram as illustrated in Figure 6. The peak is extrapolated to the baseline. It is slightly more accurate to measure the peak width at half maximum. For a symmetrically shaped (Gaussian) peak, the equation for the number of theoretical plates can be rewritten as:

$$N = 5.54 \left(\frac{t_r}{w_{1/2}}\right)^2 = \left(\frac{t_r}{w_{0.6}}\right)^2 \tag{6}$$

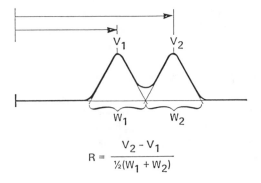

$$R = \frac{V_2 - V_1}{\frac{1}{2}(W_1 + W_2)}$$

FIGURE 7.   Resolution equation.

The number of theoretical plates is, of course, directly related to the length of the column. A longer separator column is more efficient than a short guard column, even though they may be packed with the same ion exchange material. Thus, it is more meaningful to describe the number of theoretical plates per millimeter of column length, or, as is usually done, the number of millimeters of column needed to get one theoretical plate. This is called the "height equivalent per theoretical plate" (HETP). It is simply the column length in millimeters divided by the number of theoretical plates, N:

$$\text{HETP} = \frac{L_{(mm)}}{N} \qquad (7)$$

Chromatographers also talk about the linear velocity of the eluent or mobile phase. It is simply the length of the column in millimeters divided by the time coinciding with the void volume, $t_o$.

Of primary concern to the chromatographer, however, is resolution, or "how well are two peaks separated". The measurement of resolution is illustrated in Figure 7. The retention volumes of peaks 1 and 2, $V_1$ and $V_2$, are measured, along with their widths in units of volume. Alternatively, all measurements can be made in units of time, i.e., retention times, $t_1$ and $t_2$, along with peak widths in time unit. Since $V_1$, $V_2$, $W_1$, and $W_2$ are all measured in the same units, resolution, R, is a unitless number. The resolution is directly related to the column efficiency (theoretical plates, N), the selectivity, and the capacity factor, $k'$ by the following equation:

$$R = 1/4(N)^{1/2}\left(\frac{\alpha - 1}{\alpha}\right)\left(\frac{k'}{k' + 1}\right)$$

$$R = (\text{efficiency})^{1/2}(\text{selectivity})(\text{capacity}) \qquad (8)$$

The first term, N, is related to the height equivalence per theoretical plate by HETP = L/N, where L is the column length. The factors affecting HETP were described by van Deemter[57] and are given by the van Deemter equation:

$$\text{HETP} = Au^{0.33} + \frac{B}{u} + Cu + Du \qquad (9)$$

The first term is independent of the linear velocity of the mobile phase, u. It is characteristic of the peak broadening through multiple effects. This includes the way that the analyte fits

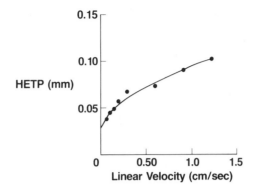

FIGURE 8.   Effect of linear velocity on HETP. The HPIC®
resin shows little change in HETP as linear velocity increases.

into the interstices of the column packing material. It depends on the size and type of packing material. Term B is characteristic of the diffusion of analyte into the pores of the column. For pellicular resins used in ion chromatography, this is negligible. Term C is a measure of the longitudinal diffusion of analyte into the mobile phase (eluent). For a more detailed description of the factors involved in the van Deemter equation see References 57 and 58. The importance of the van Deemter equation to the analyst is that the column efficiency will decrease as the linear velocity, or flow rate, of the eluent increases. For pellicular columns in which there is very little diffusion of analyte, this effect is minimal. This is illustrated for the AS-4 ("new" anion HPIC resin) in Figure 8.

## IX. PRINCIPLES OF GRADIENT ELUTION

With the advent of micromembrane suppressors and gradient pumps, gradient elution became possible for ion chromatography. Although many important separations can be accomplished with isocratic elution, they usually involve ions with similar charge. For example, the seven common anions all have −1 or −2 charges. It is more difficult to separate mono-, di-, tri-, tetra-, and pentavalent anions in one run. Similarly, if a sample contains a mixture of small, hydrophilic anions (which are weakly retained) and large, hydrophobic anions (which are strongly retained), a gradient elution may be needed.

Gradient elution was not possible with fiber suppressors because they did not have sufficient ion exchange capacity to keep up with the rapidly changing ion fluxes present in a gradient elution. The concentration of anions in the eluent would increase so fast that the fiber suppressor could not react with them to completely convert them to their low conducting form. To better describe this concept, the term "dynamic suppression capacity" is used.[20,21] It is the maximum number of equivalents of eluent that can be suppressed or neutralized per unit time. The micromembrane suppressors, with their increased dynamic suppression capacity, are capable of adequately suppressing the increased ion fluxes in a gradient elution.

To best utilize the tool of gradient elution, it is important to understand the parameters that affect the chromatographic properties of ions in a gradient. There are two major classifications of gradients in ion chromatography, corresponding to the two most important parameters. The first is the concentration of the ions in the eluent. If a gradient consists of a change in the concentration of the eluting ions, it is called a "concentration gradient". If the analyst is separating ions of the same charge, then a change in eluent concentration will not affect the selectivity of the ion chromatography. If, however, the sample contains ions of different charge, the more highly charged the ion, the more it is affected by the concentration gradient. The second type of gradient is when a new eluting species is introduced into the eluent. This is called a "compositional gradient". The composition of the gradient can be changed either by adding a new eluting ion to the eluent, or by adding another solvent to the gradient (i.e., acetonitrile in MPIC).

Some of the parameters that affect gradient ion chromatography are best understood by first examining their effect on isocratic elution. These parameters include the eluent concentration, E1, the charge on the eluent ion, E, the charge on the analyte ion, A, the ion exchange capacity of the column (resin), Q, and the ion exchange selectivity or distribution coefficient, K. The capacity factor, $k'$, of the ion, A, has been shown to be related to these parameters by the following equation:[21]

$$k' = \frac{Vr - Vm}{Vm} = \frac{Vs}{Vm} K^{1/E} Q^{A/E} (E1)^{A/E} \tag{10}$$

where $V_r$ is the retention volume of the analyte ion, $V_m$ is the column void volume, and $V_s$ is the volume of the stationary phase. The distribution coefficient, K, the ion exchange capacity, Q, and the charges of the eluent and analyte ions (E and A) do not change when using a gradient elution. Therefore, these terms can be combined into a single isocratic column constant, $C_i$. Thus:

$$C_i = \frac{Vs}{Vm} K^{1/E} Q^{A/E} (E1)^{A/E} \tag{11}$$

and $k'$ can be redefined as:

$$k' = C_i (E1)^{-A/E} \tag{12}$$

Taking the log of both sides:

$$\log k' = \log C_i - \frac{A}{E} \log (E1) \tag{13}$$

The analyst can then determine the retention time (and calculate $k'$) for an analyte ion in an isocratic elution. The $k'$ for this ion can then be measured again, using a different concentration of eluent, $E_1$.

This series of experiments can be performed at several isocratic eluent concentrations. A plot of the measured log $k'$ values vs. the eluent concentration will be a straight line with a negative slope equal to A/E, the ratio of the charge of the analyte to the charge of the eluent ion. Such plots are shown in Figure 9 for chloride, fumarate, nitrate, sulfate, and citrate, determined through AS-6 column analysis.

Recently, a separator column very similar to the AS-5 has been introduced. The major difference between the AS-5 and the new, AS-5A column is that the AS-5A column has smaller polystyrene/divinylbenzene particle size. For the AS-5 it is 10 μm and for the AS-5A it is 5 μm. This smaller particle size produces a more efficient column. Thus, the AS-5A has been called the gradient column for anions.

When this AS-5A column is used to evaluate the slopes of plots such as those shown in Figure 9, the results agree quite well with theory. For the monovalent anions chloride and nitrate, Equation 13 predicts a slope of −1.0. The experimentally determined slopes were 1.03 and 0.95. For the divalent anions, sulfate and fumarate, the predicted slope is −2.0. The observed slopes using the AS-5A are −2.10 and −2.03. These results are summarized in Table 1.

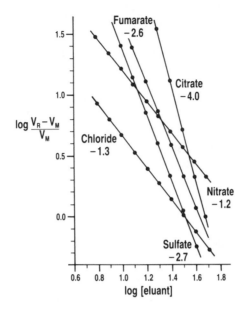

FIGURE 9.   Retention vs. eluant concentration. Isocratic elution on AS6 with 4-cyano-phenol eluant.

## Table 1
## ANIONS DETERMINED BY AS-5A COLUMN ANALYSIS

|  | Chloride | Nitrate | Sulfate | Fumarate | Citrate |
|---|---|---|---|---|---|
| Isocratic |  |  |  |  |  |
| Slope | − 1.03 | − 0.95 | − 2.10 | − 2.03 | − 3.58 |
| − A/E | − 1 | − 1 | − 2 | − 2 | − 3 |
| Const$_i$[a] | 27 | 110 | 3900 | 7800 | 9800000 |
| Gradient |  |  |  |  |  |
| Slope | − 0.50 | − 0.47 | − 0.67 | − 0.65 | − 0.77 |
| − A/(A + E) | − 0.5 | − 0.5 | − 0.67 | − 0.67 | − 0.75 |
| Cons$_g$[b] | 7.4 | 15 | 20 | 26 | 42 |
| Const$_g$(calc)[c] | 7.0 | 14 | 19 | 24 | 65 |

[a]   Inverse log of y intercept from Figure 4. Represents k′ for a 1-m$M$ NaOH eluant.
[b]   Inverse log of y intercept from Figure 5. Represents $(V_r–V_m)/V_m$ for gradient elution with a ramp-slope of 1 mM/m$\ell$.
[c]   Calculated from Const$_i$ using Equation 8.

The important result is that the relative retention of analytes can be predicted based on Equation 13. Overall trends are observed that can be used in predicting the best gradient program to use. For analytes of the same charge as the eluent ion, a change in the eluent concentration does not change the selectivity (because their slopes are the same). For a monovalent analyte ion with a monovalent eluent, doubling the eluent concentration decreases the capacity factor, k′, in half. For a divalent analyte ion with a monovalent eluent, doubling the eluent concentration would cut the k′ to one fourth its original value. The fact that the slopes vary for ions with different charges indicates that a variety of selectivities can be obtained by gradient elution. Eluent concentration has a dramatic effect upon ion exchange selectivity when separating ions of different valency. For example, when separating phosphates with charges from − 1 to − 6, i.e., phosphate through tetrapolyphosphate, the task was impossible with isocratic elution and conductivity detection using low-capacity fiber suppressors. With the advent of the micromembrane suppressor and gradient ion chromatography, the very strongly retained $k'$ of the tetrapolyphosphate decrease by a factor of 16. The major accomplishment of gradient elution is to move together peaks that are very far apart in an isocratic elution. This is especially true when the peaks are due to ions with different charge.

In gradient elution, though, $k'$ varies as a function of eluent concentration. As the eluent concentration increases, the observed retention time decreases. The concept of $k'$ is slightly different in a gradient, however. It is perhaps better to consider $k'$ as being an instantaneous capacity factor, and rename it $k_a'$. Similarly, there is no fixed eluent concentration, $E_1$, but instead an instantaneous $E_1$, which is given by:

$$[E_1] = RV \tag{14}$$

where R is the gradient ramp rate, or the rate at which the gradient changes in units of millimolar per milliliter, and V is the volume of eluent pumped since the beginning of the gradient.

The instantaneous capacity factor, $k_a'$, can be calculated by substituting Equation 14 into Equation 15, i.e.,:

$$k'_a = Const_i(RV)^{-A/E} \tag{15}$$

In other words, $k_a'$ is the instantaneous capacity factor that would result if the eluent concentration at that instant had actually been held constant from injection to elution. Instead, $k_a'$ is an "effective capacity factor". To calculate this effective capacity factor, it is necessary to integrate $k_a'$ from injection to elution. To do this, consider $k_a'$ as $dV/dx$. A fractional volume of the eluent, $dV$, passes over the band maximum and moves the band a fractional distance, $dx$, down the column. Thus, Equation 16 can be rewritten as:

$$k'_a = \frac{dV}{dx} = Const_i(RV)^{-A/E} \tag{16}$$

which can be rearranged and integrated:

$$\int_0^{V_r - V_m} Const_I^{-1}(RV)^{A/E}dV = \int_0^{V_m} dx \tag{17}$$

$$\frac{V_r - V_m}{V_m} = \left(\frac{E}{A + E}\right)^{\frac{-E}{A+E}} V_m^{\frac{-A}{A+E}} Const_i^{\frac{E}{A+E}} R^{\frac{-A}{A+E}} \tag{18}$$

The three constants in front of the R term can then be combined into one column gradient constant, $Const_g$. Equation 19 can then be rewritten as:

$$\frac{V_r - V_m}{V_m} = Const_g R^{\frac{-A}{A+E}} \tag{19}$$

Taking the log of both sides:

$$\log \frac{V_r - V_m}{V_m} = \frac{-A}{A + E} \log R + \log Const_g \tag{20}$$

Again, R has units of millimolar per milliliter, and represents the gradient ramp rate. Equation

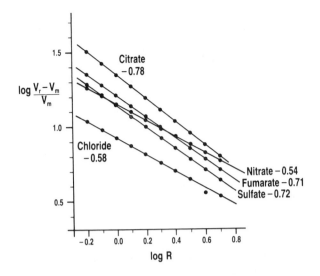

FIGURE 10.   Retention vs. ramp rate of gradient. Gradient is on an AS6 column using 4-cyanophenol eluant.

20 predicts that a plot of $\log(V_r - V_m)/V_m$ vs. log R will be linear for each analyte. Thus, the analyst can measure the retention time of each analyte at a given ramp rate. This experiment can be repeated for several ramp rates and be plotted according to Equation 20. The slopes are predicted to be negative and equal to (A/A + E). For a monovalent eluent such as $p$-cyanophenate, the predicted slope would be $-1/2$. For di- and trivalent analytes, the slopes would be $-2/3$ and $-3/4$. The results of such experiments[21] are shown in Figure 10. The results from isocratic experiments can be used to help interpret the data from these gradient experiments. The effective eluent charge, E, can be taken from the slope of the isocratic plot. In other words, when performing the preliminary isocratic experiments, log $k'$ was plotted against the log $E_1$ (the eluent concentration). The slope was negative and equal to A/E. Assuming A to be accurate, the effective E (eluent charge) is simply equal to A/slope. Similarly, $C_i$, the isocratic column constant, is equal to the y-intercept of the isocratic plot. Remembering that the column constant for gradient elution, $Const_g$, is equal to the first three terms in Equation 18, $Const_g$ can also be calculated. This can be compared to a $Const_g$ that is measured by the y-intercept of the gradient plot (as predicted by Equation 20, and illustrated in Figure 10. A comparison of (A/A + E) to the gradient slope was shown in Table 1.

Again, E is the apparent eluent charge calculated from the slope of the isocratic plot, and Equation 14 predicts that the gradient slope will be negative and equal to (A/A + E). These predictions are fairly well substantiated by the data.

Similarly, the column gradient constant, $Const_g$, can be calculated from Ci, R, and $V_o$. The calculated $Const_g$ agrees with the $Const_g$ determined from the y-intercept of the gradient plot as shown in Table 1.

As described by Rocklin et al.,[21] the optimum conditions for gradient elution are theoretically given in Equation 20. There are practical limitations, however. Equation 20 assumes that the initial eluent concentration is zero. Faster gradient runs can be obtained by starting the gradient at a higher eluent concentration. It is sometimes useful to combine several gradient ramp rates and to even include some isocratic steps. Such complicated gradient programs would alter Equation 20 to make it unwieldy. In spite of this, Equation 20 can be used to predict trends. If two ions of different charge co-elute, increasing the gradient ramp rate will cause the ion with higher charge to elute first. This is demonstrated in Figure

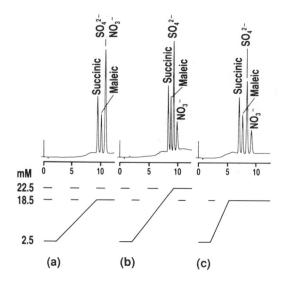

FIGURE 11.    Effect of gradient ramp on elution on AS6, using
4-cyanophenol. Divalent ions are more strongly affected by
increasing eluant concentration than are monovalent ions.

11, where sulfate and nitrate co-elute in the first gradient. The initial concentration of *p*-cyanophenate eluent is 2.5 m*M* in all three chromatograms. The concentration is increased to 18.5 m*M* between 2 and 7 min after injection in the first chromatogram. In the second, the concentration is increased to 22.5 m*M* in the same 2 to 7 min. This results in baseline resolution of nitrate and sulfate. In the third chromatogram, the gradient is from 2.5 to 18.5 m*M,* but in 5 instead of 9 min (as in the first). This causes the divalent anions to elute earlier, but because of the long period of holding at 18.5 m*M,* the resolution of the divalent anions is better than in the second chromatogram.

It might be noted, by those who have seen chromatograms of the seven common anions before the advent of gradient elution, that nitrate always eluted before sulfate. It is important to remember that much lower eluent concentrations were used in the isocratic runs. The retention of divalent ions is reduced much faster by an increased eluent concentration than is the retentin of monovalents. This is predicted in Equation 20.

Thus, the optimization of a gradient program for ion chromatography should start with isocratic data and the predictions based on Equations 12 and 13. However, some experimentation is needed to determine the best initial eluent concentration and the optimum combination of ramp rates and isocratic steps. This has been done on the AS-6 with *p*-cyanophenol eluent[21] as shown in Figure 12. This illustrates the separation of 22 anions in 22 min. This would require three isocratic runs if separated individually.

Before optimizing a gradient program, it is desirable to select an eluent. Prior to gradient elution, the vast majority of anion separations used a bicarbonate/carbonate eluent. This eluent is not so useful in gradients. It is important to use eluents which can be suppressed to produce almost constant background conductivities on the order of a few microSiemens. Salts of weak acids are potential eluents in gradient ion chromatography. The weakest acid in aqueous solution is water itself, with the salt being a strong base such as NaOH. It is an excellent choice as a gradient eluent because it is converted in the anion micromembrane suppressor to water, regardless of its concentration. As long as the capacity of the suppressor is not exceeded, the eluent hydroxide has no direct effect on the background conductivity. The dynamic suppression capacity of an anion micromembrane suppressor is usually sufficient to suppress 100 m*M* NaOH flowing at 1.0 m$\ell$/min. An example of a gradient elution

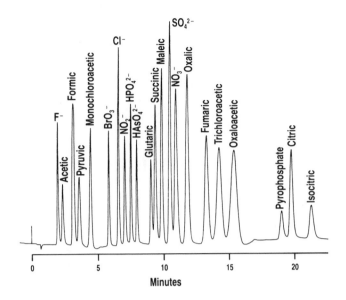

FIGURE 12.   Gradient elution of anions.

using NaOH was shown in Figure 4 in which the charges on the analytes varies from $-1$ (chloride) to $-6$ (tetrapolyphosphate).

There are other salts of weak acids that can be suppressed to give low conductivities. This includes borate and $p$-cyanophenate. In general, the salts of acids with pKa $>7$, are potential eluents. As pKa decreases, the extent of dissociation increases, resulting in higher background conductivity. On the other hand, salts of acids with pKa as low as 5 can be used for isocratic elution because the background conductivity will remain constant.

The other type of eluent that can be used in gradient elution is an anion that will produce a zwitterion following suppression. Amino acids such as glycine are examples. The closer the isoelectric point of the eluent ion is to 7, the better gradient eluent it would make.

Another consideration in selecting an eluent is to consider the charge on the eluent ion. Monovalent eluents have an advantage over divalents because a given change in eluent concentration produces a greater change in analyte retention than divalent eluents as predicted by Equation 20. Doubling the concentration of a monovalent eluent will cause $k'$ for a monovalent analyte to be decreased to 50% of its initial value. A similar doubling of a divalent eluent concentration will only decrease the $k'$ of a divalent analyte by a mere 25%. If the eluent was divalent, doubling its concentration would decrease the $k'$ of a monovalent to 71% and the $k'$ of a divalent analyte to 50% of its original value. Compared to a monovalent eluent, the concentration of a divalent eluent must be increased much more to elute ions of very different retention.

One of the problems associated with gradients in liquid chromatography is that contaminants in the eluent can produce spurious peaks in the chromatograph. A gradient begins with a weak eluent in which most analytes would tend to be retained on the column. As the eluent strength increases during the gradient, those analytes which were strongly retained in the weak eluent may now elute. This produces spurious peaks in the chromatogram. In gradient HPLC, this problem was overcome by the development of high-purity solvents. These solvents have so few contaminants that there are essentially no analyte contaminants available to cause spurious peaks. Similarly, in gradient ion chromatography, it is necessary to purify the eluent. For example, in the NaOH gradient used in Figure 12, sulfate would be strongly retained at the beginning NaOH concentration of 2.5 m$M$. As the gradient continues, more sulfate-contaminated eluent would flow past the AS-6 and more sulfate

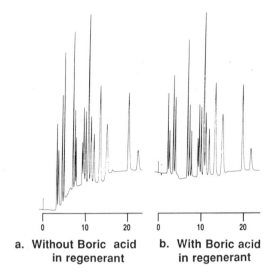

a. **Without Boric acid**
**in regenerant**

b. **With Boric acid**
**in regenerant**

FIGURE 13.    Balancing baseline with mannitol and boric acid using an AS6 column and 4-cyanophenol eluant. The boric acid forms an ionic complex with mannitol, which raises the background conductivity at the beginning of the gradient. As the gradient increases, the boric acid mannitol is decreased to keep the background conductivity constant.

would be strongly retained. Once the NaOH concentration becomes high enough, the previously strongly retained sulfate would elute. This would produce a peak that would incorrectly give the analyst the impression that there was really something in the sample that was actually a contaminant in the eluent. To overcome this problem, a high-capacity, low-efficiency anion exchange column is placed before the load/inject valve. In this way, a significant amount of eluent could pass through this column and bind the sulfate. If and when the sulfate ever does elute, the low efficiency of the column would cause the peak to be so broad as to make it undetectable except as a small baseline drift.

These and other baseline drifts can be minimized. Actually, the major cause of baseline drifts is the increase in eluent concentration during the gradient. When NaOH is the eluent, this is already minimized because the eluent is being suppressed to water regardless of the NaOH concentration of the gradient. If glycine was being used in the gradient, it would be suppressed to its zwitterion form, which has a small but finite conductivity. As the glycine concentration increases, so does the small but measureable conductivity of the background.

There are two ways of balancing the baseline. One is to run the gradient as a blank, i.e., run the gradient without injecting a sample, and store the result on a computer. This baseline blank can then be substracted from chromatograms run using the same gradient. The other way is to balance the baseline chemically. This can be done by adding mannitol to the weak eluent and boric acid to the regenerant. Boric acid forms a complex with mannitol which is a stronger acid than either of the two by themselves.[59] Boric acid in the acid regenerant is neutral, so it can diffuse through the cation exchange membranes of the micromembrane suppressor. It can then combine with the mannitol in the suppressor, producing a conductive species. By decreasing the mannitol concentration at about the same rate that the concentration of the eluting anion increases, the baseline can be balanced. The gradual increase in background conductivity due to the increase in eluent concentration can be offset by the decrease in conductivity due to the decrease in concentration of the boric acid-mannitol. The effects of this chemical baseline balancing is illustrated in Figure 13. Mannitol and HCl were used in the second chromatogram, but not in the first.

Another application of gradients in ion chromatography is in MPIC separations. Reversed-phase HPLC is similar to MPIC, so experience from gradients in HPLC can be useful. As

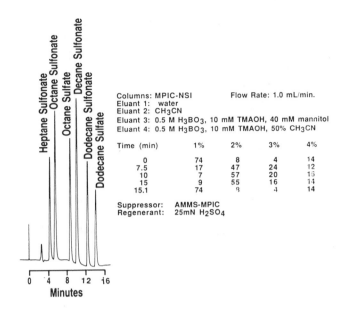

FIGURE 14. Gradient ion pairing. Column: MPIC, detection: chemically suppressed conductivity.

with ion pair chromatography in reversed-phase HPLC, the gradient in MPIC may be either in the percent of organic solvent or in the eluent. The ionic strength of the eluent may not change when only the percent organic solvent changes. However, the background conductivity may change as the dielectric constant of the eluent changes. As a rule, the percent organic solvent will increase during the gradient, so the dielectric constant and the conductivity background will decrease. This baseline change can be corrected by either of the two procedures discussed, or it may be overcome by simply adding small amounts of very weakly retained anions to the eluent during the gradient. An example of gradient MPIC is shown in Figure 14, where five alkyl sulfates and sulfonates are resolved.

Thus, gradient elution in ion chromatography is already proving to be a powerful analytical technique. It was made possible by the introduction of micromembrane suppressors with high dynamic suppression capacity. The increasing concentration of eluent produces only a limited increase in background conductivity. Perhaps the best eluent for anion gradient elution is sodium hydroxide, which is suppressed to water, regardless of the concentration of the NaOH. Gradient elution is especially useful in bringing together peaks that are spread widely apart in an isocratic elution. Typically, polyvalent anions are retained longer on an anion separator column as compared to monovalent anions. In a gradient elution, the higher the charge on the anion, the more it is affected by the increasing eluent concentration. As a result, the retention time of the polyvalent anions decreases rapidly as the eluent concentration increases, but the retention time of the monovalent anions decrease much slower, as predicted by Equation 21. In an isocratic elution, the peaks due to polyvalent anions are widely separated from monovalents, often so much so that the polyvalents cannot be eluted using an eluent strength necessary to separate monovalents. The gradient elution brings these widely separated peaks closer together. As a result, more anions (or cations) can be separated in one run.

## X. CONDUCTIVITY

To properly develop analytical methods using chemically suppressed conductivity, it is important to understand some basic principles that apply to conductivity. Such as understanding will enable the analyst to anticipate the shape of calibration curves. Because of the physics behind electrical conductance, along with the nature of the ion-exchange separations, it is expected that nonlinear regions will exist in plots of conductivity vs. ion concentration.

There are at least three terms that are used to describe the electrical conductance of ions in solution. Specific conductance, $\kappa$, is the conductance across the opposite faces of a cube of solution, 1 cm on an edge. The equivalent conductance, $\Lambda$ is the conductance of 1 equivalent of completely dissolved electrolyte, dissolved in 1 m$\ell$ volume. The limiting equivalent conductance is the equivalent conductance of an ion at infinite dilution. It is determined experimentally be measuring the conductance of the electrolyte at different concentrations. The measured conductances are plotted against the concentration of electrolyte. The plot is extrapolated to infinite dilution. The conductance at this infinite dilution is the limiting equivalent conductance.

The equation relating specific conductance to equivalent conductance is

$$\kappa = \frac{\Lambda N}{1000} \tag{21}$$

where N is the normality of the solute and is defined as the total concentration of positive and negative charges produced upon dissociation. For an electrolyte $M_x^{m+}X_m^{x-}$, $N = m \times C$, i.e., for $Na_3PO_4$:

$$N = 3 \cdot 3 \cdot C \tag{22}$$

where C is the molar concentration.

A conductivity detector is measuring the total conductance of the solution that flows through it. The electrical conductance of any solution flowing through the detector at a given instant depends on the type and concentration of the ions in the solution at that instant. The equivalent conductance for NaCl, for example, is equal to the sum of the equivalent conductances of the $Na^+$ and $Cl^-$, i.e.,

$$
\begin{array}{ccccc}
\text{Salt} & & \text{Cation} & & \text{Anion} \\
\Lambda & + & \lambda & + & \lambda \\
M_x^{m+}X_m^{x-} & & M_x^{m+} & & X_m^{x-}
\end{array} \tag{23}
$$

This equation is a mathematical interpretation of the law of independent migration first described by Kohlrausch. In a dilute solution, as is seen in ion chromatography, the specific conductance is equal to the sum over all ions of the equivalent conductances of each ion multiplied by its normality. The conductivity depends on both the anion and cation conductances. The limiting equivalent conductances of most ions are between 30 to 80 $\mu$S/cm. Two notable exceptions are solvated proton and hydroxide, with limiting equivalent conductances of 350 ($H_3O^+$) and 192 ($OH^-$). The exceptional conductances of these ions are due to their high mobility in solution. This mobility is caused by a unique transport mechanism for hydronium and hydroxide. They can "move" by a shift in hydrogen bonds in the water structure which produces a charge transfer to a location distant from the original ion. There is mobility of charge without direct movement of a particular ion.

The high conductances of $H^+$ (more properly $H_3O^+$) and $OH^-$ is the basis for the enhancement of analyte conductivity in chemically suppressed conductivity. For example, it was already mentioned how ions such as sodium and chloride have limiting equivalent conductances between 30 and 80 μS/cm. The anion suppressor column will exchange H for sodium in the sample to form the more highly conducting electrolyte, HCl. Similarly, the cation suppressor will exchange $OH^-$ for $Cl^-$ in the sample to form the more highly conducting NaOH. It would be an oversimplification to think that the conductance measured for an ion would be a perfectly linear function of concentration for all concentrations of analyte; even in theory, this is not true. At very low and at very high concentration, there is some nonlinearity expected, due to the basic properties of ions in solution. The most significant deviation from linearity occurs at higher analytye concentrations, so the reasons behind this, will be discussed first.

This deviation from linearity begins at different concentrations for different ions. For weakly dissociated species, the linearity range will depend on the degree of dissociation of the ion. For example, ammonia, with a pKa = 9.2, will be fully dissociated at milligrams per liter (parts per million) concentrations, but will not be fully dissociated at higher concentrations. The degree of dissociation, and therefore the conductance, will depend on the ammonia concentration. The weaker the acid or base, the sooner the calibration data will become nonlinear.

In addition, the presence of other ions will affect the degree of dissociation of a weak acid or base. Addition of a common ion will increase the degree of dissociation. For example, the addition of NaOH to $NH_4OH$ will increase the degree of dissociation of this weak base, just as the addition of HCl will increase the degree of dissociation of $H_3PO_4$. This is important to remember when analyzing samples with high ionic strength such as brines and plating solutions. In these analyses, the standards should be prepared in solutions that mimic the sample as closely as possible. For example, when determining trace chloride in chromic acid, sulfuric acid, or NaOH, it is important to prepare standards in high-purity chromic acid, sulfuric acid, or NaOH at the same concentrations expected to be present in the sample.

The presence of too many ions of even strong acids or bases in solution, will also cause nonlinearity in plots of conductance vs. concentration. This is due to ion-ion interactions. As a result of electrostatic attractions between ions of opposite charge in concentrated solution, ion pairs can form. Several electrolytes may be completely ionized, but not completely dissociated. Ion pairs in such concentrated solutions do not have as high equivalent conductances as the free ions would have. Thus, the conductance is no longer a linear function of concentration. In dilute solutions (millimolar concentrations or lower), ion-ion interactions become negligible and solutes are both completely ionized and dissociated. Ion-pair formation is for small and high valence ions in a low dielectric strength solution. The ionic size can be considered to be the effective size of the ion in solution. It includes the ion itself and its solvation sphere. The solvation sphere is on the order of 1 Å. It starts to decrease as the concentration increases and the charge (or valency) of the ion increases.

In the concentration ranges used in ion chromatography, the ion-pair formation is negligible. Usually a technique such as ion chromatography is evaluated partly by its sensitivity. The high sensitivity of chemically suppressed ion chromatography means that detection limits in the range of $5.10^{-12}$ μg (i.e., 0.1 ppb of $Cl^-$ in a 50-μℓ injection loop) can be obtained. Certainly, ion-ion interactions are negligible at concentrations well above this. Since ion chromatography is usually thought of as a method for trace analysis, one does not usually consider the possibility of nonlinear calibration curves. Ion chromatography does also have good utility in analyzing concentrated solutions of ions as well. The question may then arise about how far to dilute the sample. The answer is to dilute the sample so that ion-pair formation becomes negligible, and weak acids (or bases) are fully dissociated.

The decrease in equivalent conductance at high concentrations is expressed in the Debye-Huckel-Onsager equation:

$$\Lambda = \Lambda_0 - (A + B\Lambda_0)\sqrt{C} \qquad (24)$$

This equation assumes complete dissociation of the electrolyte and attributes a decrease in conductance to the decrease in ionic velocity resulting from ionic forces. These ionic forces exert an electropotential on the neighboring ions. The constant A is dependent on the dielectric constant and viscosity of the solvent. It also depends on the temperature and it accounts for electrophoretic retardation in the mechanism of transport. This may be explained by the fact that under the influence of the applied potential the central ion moves in the direction opposite to its solvation sphere (they have opposite charges). This exerts a retarding force on the ionic migration. Coefficient B accounts for the asymmetry, or relaxation effect. During the movement of the ion, the charge on the solvation sphere surrounding it becomes asymmetrical. For a finite period of time, charge density is greater behind the moving ion than in front of it. This effect causes retardation of the ion movement under the influence of the applied potential. This, in turn, causes a decrease in conductance.

It should be noted that this C dependence is only applicable for concentration ranges higher than those typically analyzed by chemically suppressed ion chromatography. In the parts per million range of concentrations, the linear dependence of conductance on concentration is described by the Kohlrausch Law (Equation 24).

Another factor affecting the degree of dissociation of ions is the solvent composition. This is especially important in MPIC. The nature of the solvent can play an important part in controlling solute dissociation, and will, in turn, affect solute conductance. The addition of an organic solvent to water will lower the eluent conductivity. As the dielectric constant decreases, the higher the electrostatic forces between the ions, and the lower the dissociation. It is important to remember that if organic solvent is added to the eluent it will affect conductivity detector response for both weakly and strongly dissociated species.

Temperature will also affect conductivity. An increase in temperature causes ion conductance to increase at infinite dilution in accordance with this equation:

$$\lambda^{\circ}_{t} = \lambda^{\circ}_{25} \, e^{k(t - 25)} \tag{25}$$

where $\lambda^{\circ}_{t}$ $\lambda^{\circ}_{25}$ are the equivalent conductances at infinite dilution at temperature T, and at 25°C, respectively. The factor k is constant for a given ion and solvent. For most ions in water, k is about 0.02 at 25°C. Thus, the equivalent conductance of most ions increases by approximately 2% per degree Celsius. This can be compensated for in an ion chromatograph by setting the temperature adjustment to 2% per degree Celsius.

A more detailed description of conductance in ion chromatography was given by Doury-Berthod et al.[60] They developed a general expression for conductance. Their equations can be made more specific for the case of phosphate as the analyte and bicarbonate/carbonate as the eluent. Before the phosphate-containing sample is injected on an anion separator, the column is equilibrated with eluent. In other words, bicarbonate and carbonate are both bound to the column. As the phoshate reaches the top of the separator, it exchanges with the bound $HCO_3^-$ and $CO_3^{2-}$. The following equilibria are established:

$$HPO_4^{2-} + \overline{2HCO_3^-} \rightleftarrows \overline{HPO_4^{2-}} + 2HCO_3^- \tag{26}$$

$$HPO_4^{2-} + \overline{CO_3^{2-}} \rightleftarrows \overline{HPO_4^{2-}} + CO_3^{2-} \tag{27}$$

where bars represent the stationary phase (i.e., bound ion). If the sample had been diluted into the same bicarbonate/carbonate used as the eluent, there was a local increase in the total concentration of the $HCO_3^-$ and $CO_3^{2-}$ in the mobile phase (i.e., the exchanged bicarbonate and carbonate enter the mobile phase which already contains some $HCO_3^-$ and $CO_3^{2-}$). There is also a decrease in the concentration of $HCO_3^-$ and $CO_3^{2-}$ in the stationary phase. This excess of $HCO_3^-$ and $CO_3^{2-}$ in the mobile phase is accompanied by an equivalent cation ($Na^+$) excess. Because it has a lower affinity for the separator column, the $HCO_3^-$ and $CO_3^{2-}$ move through the column faster than does the analyte, $HPO_4^{2-}$. This produces a positive peak before the peak due to $HPO_4^{2-}$. Just as there is a local increase of bicarbonate and carbonate eluting before the phosphate, there is a local decrease of $HCO_3^-$ and $CO_3^{2-}$ when $HPO_4^-$ elutes. Since the suppressor column adds protons to the phosphate, bicarbonate, and carbonate, it can be said that there is a deficiency of $H_2CO_3$ reaching the conductivity detector when $H_3PO_4$ reaches the detector. When the phosphate concentration is above 0.1 m$M$, the deficiency of weakly conducting $H_2CO_3$ is negligible. At very low phosphate (or other analyte) concentrations, though, the equations developed by Doury-Berthod predict a slight nonlinearity in the calibration.

The total conductance, G, of the solution which flows through the detector cell is the sum of the contributions from $H^+$, $HCO_3^-$, $H_2CO_3$, $H_2PO_4^-$, and $HPO_4^{2-}$, i.e.,:

$$G = G1 + G2 + G3 \qquad (28)$$

In other words, the change in conductance from the background is a combination of contributions from G1, G2, and G3. G1 is due to $HPO_4^{2-}$ and $H^+$. G2 contains contributions from $HCO_3^-$ and $H^+$ from the eluent. It is due to the local decrease in $HCO_3^-$ bound to the separator. G3 is the background conductance of the eluent. Based on this equation, the theoretical, calculated plot of conductance vs. concentration at very low concentrations, becomes slightly nonlinear.

## XI. COLUMN CARE

Chromatographers have been plagued throughout the years with gradual deterioration of column performance. This is especially true when analyzing samples that have a complex matrix. Soil scientists may want to determine nitrate, phosphate, and sulfate in soil that contains the ill-defined humic acids. Toxicologists may want to determine oxalate in urine that contains "urochromes". Electroplaters may want to determine chloride and sulfate in plating solutions that have unknown organic additives. The humic acids, urochromes, and organic additives can foul a separator column. This deterioration of column performance can be seen by a loss of resolution, increased operating pressures, discoloration of the column, abnormal peak shapes, or column voids.

There are some obvious ways to minimize column fouling. One way is to know as much about the sample matrix as possible, and avoid injecting anything harmful on a column. This is relatively easy for the analyst who performs ion chromatography on highly purified deionized water, and is detecting parts per billion levels of anions. There are very few substances that will cause deterioration of column performance when present at parts per billion levels. A corollary to this is that when analyzing samples that are in a complex, unknown matrix, dilute the sample as much as possible. Instead of using a dilution that has the desired peaks on scale at the 30-μS scale, dilute the sample another 10- to 30-fold so that the 3- or 1-μS scale can be used. The baseline may not look so nearly perfect, like it does on a 30-μS scale, but the separator columns will survive longer. Another easy way to minimize column deterioration is to always filter the samples before injecting. Fortunately, this is done automatically when using a modern autosampler. Thus, automation not only provides a much larger productivity, but also is useful in minimizing column deterioration.

Probably the most obvious way to minimize column deterioration is to use a guard column to stop harmful humic acids, urochromes, or organics from ever reaching the separator column. The guard column is simply a short separator column. This means that it can be cleaned faster once it is fouled, and that it can be replaced for a lower cost if that becomes necessary. It is important to know when a guard column should be cleaned or replaced. If the guard column is not cleaned or replaced, it can become saturated with the harmful compounds or ions. At this point, the harmful substances will reach the separator columns and will begin to deteriorate its performance. Thus, it is useful to monitor the guard column in a systematic way. This is rather easy if a single guard and separator column are to be used repetitively for one analysis. The separator column should first be removed so that only the guard column and suppressor column are installed. Next, a standard containing a weakly retained ion and an ion that has a $k'$ of about 1.3 to 1.4 should be injected. For the anion separator columns, a mixture of fluoride and sulfate would be useful. The fluoride provides a good measure of the column void volume, and the sulfate elutes after about three void volumes. The $k'$ for sulfate is measured. The actual sample to be analyzed routinely is then injected five to ten times. The fluoride plus sulfate standard is then re-injected and $k'$ remeasured for sulfate. If the samples were quite clean, it is possible that no change in $k'$ will be observed. If this is the case, these samples can be analyzed for a few days or a few weeks. At some time, though, the separator column should be removed and the fluoride plus sulfate reinjected and $k'$ remeasured for sulfate. This should be repeated periodically. The results can be plotted on a graph as $k'$ vs. number of injections. The plot should be extrapolated to $k'$ of 1/2 of the original $k'$ (i.e., after the first injection of standard). There will come a point at which the $k'$ of the indicating ion (sulfate in our example) decreases to 50% of its original value. At this point, the guard column should be cleaned. This may happen after 100 injections or after 1000 injections, but it is important to know when the guard column or precolumn needs to be cleaned.

When it becomes necessary to clean a guard column, it is useful to know what the probable cause of contamination could be. In general, if samples contain strongly retained species, it is best to try a more concentrated eluent to clean them. For example, if samples are analyzed using a 3.0 m$M$ bicarbonate/2.4 mM carbonate on an AS-4A, strongly retained anions may be effectively removed by washing the guard column (AG-4A) with 0.1 $M$ sodium carbonate. When washing any guard column with any solution, though, it is very important first to remove the separator column. This is important because 0.1 $M$ sodium carbonate may effectively elute some large, hydrophobic anions off the short guard column that has a limited ion exchange capacity. The longer separator column, however, has a higher total ion exchange capacity. It is quite possible that a contaminating anion could be washed off the AG-4A, but then be strongly retained on the AS-5A. Thus, if the separator column is not removed, it is possible that the separator column could be fouled in the process of cleaning the guard column. Similarly, it may be a good idea to also by-pass the suppressor column, The suppressor column is very rugged and probably cannot be fouled, but is serves no purpose in cleaning a guard column. At any rate, it is quite easy to simply disconnect all columns downstream from the guard column during cleaning operations.

When deciding what washing solution to use in cleaning a guard column, it is important to know what compounds or ions can foul the different types of columns. Anion guard (and separator) columns can be fouled by strongly retained inorganic anions. For example, citrate, molybdate, and chromate are strongly retained on the AG-1, 2, 3, 4, and 4A columns. Metal cyanides are strongly retained on the AG-5. Anion guard columns can also be contaminated by metals. Although the anion guard (and separator) columns contain anion exchange latex beads, these beads do not completely cover the sulfonated polystrene/divinylbenzene. There are residual cation exchange sites ($-SO_3$) on the anion guard columns, and these residual sites can bind metals. Organics, particularly those containing aromatic rings, can foul anion guard (and separator) columns. The phenyl groups interact strongly with the polystyrene/divinylbenzene core of the guard column. Aliphatic organics, especially surfactants, can also interact strongly with the polystyrene/divinylbenzene and contaminate the anion guard column.

Cation guard columns are also readily contaminated by organics. Aromatic amines and several aliphatic amines (expecially di- and triamines) are strongly retained. When using the CG-1 or CS-1, it was not possible to simultaneously separate mono- and divalent cationsin one run. Thus, divalent cations (alkaline earths) are not eluted with monovalent cations, and can contaminate the CG-1. This is not such a problem with the CG-3 (or CS-3) in which mono- and divalents can be separated in one run.

ICE separator columns are similar to cation guard columns. Because they have cation exchange sites, they are readily contaminated by cations and metals. Because ICE columns are porous, they are especially susceptible to contamination by aromatic organics. The phenyl groups on the organic column interact strongly with the polystyrene/divinylbenze of the ICE column. Because of the porosity of the ICE columns, there is a substantially larger exposed surface area of polystyrene/divinylbenzene.

MPIC guard and separator columns, having no fixed ion exchange sites, are not contaminated by metals, cations, or anions. The MPIC column can be contaminated by nonionic organics, however.

The clean-up procedure depends on the suspected contaminants and the column which is contaminated. To clean inorganic anions off an anion guard column, by-pass the separator and suppressor, rinse with 20 m$\ell$ deionized water, then wash with 20 m$\ell$ of 0.1 $M$ sodium carbonate. After the washing, re-equilibrate with the desired eluent and inject the fluoride plus sulfate standard and measure the k' for sulfate. If the k' for sulfate is not increased substantially, rinse the guard column with another 20 m$\ell$ deionized water and wash with 20 m$\ell$ 0.1 $M$ sodium sulfate. The sulfate has a higher affinity for the anion exchange sites on the resin and is more likely to wash off very strongly retained anions. Next, wash-off the sulfate by rinsing with 30 m$\ell$ 0.1 $M$ sodium carbonate. Reequilibrate with the desired eluent and remeasure the k' for sulfate.

If metallic contaminants are suspected, again by-pass the separator and suppressor and start with 20 m$\ell$ deionized water wash. The metals can then be removed by washing with 30 m$\ell$ of sodium tartarate. Tartarate is a good chelating agent for most metals. Finally, equilibrate with the desired eluent and remeasure k' for sulfate.

To remove organic contaminants, by-pass the separator and suppressor. Pump 30 m$\ell$ acetonitrile or methanol through the anion guard column. The concentration of acetonitrile or methanol should not exceed 10% for the AG-4 through AG-7 (or AS-4 through AS-7). It is also important not to expose the anion fiber suppressor (AFS-1) to organic solvents (although the micromembrane suppressor is quite stable to acetonitrile and methanol). Again, the final step is to reequilibrate with the desired eluent and remeasure k' for sulfate.

For cation guard columns, a standard consisting of lithium and potassium would be useful. If alkaline earth metal contaminants are suspect on an AG-1, by-pass the separator and suppressor and pump 20 m$\ell$ of 1.0 $M$ HCl through the guard column. Equilibrate with the desired eluent, and remeasure k' for potassium (use the retention time of lithium as an estimate of void volume). If metallic contaminants are suspect, first wash with 20 m$\ell$ deionized water, then 20 m$\ell$ of 0.1 $M$ sodium tartarate. Flush with 30 m$\ell$ of 1.0 $M$ HCl to reconvert the resin in the cation guard column from the sodium form (due to sodium tartarate) to the hydrogen form (due to HCl). Again, remeasure the k' of potassium. If organic contaminants are feared, pump acetonitrile or methanol (at concentrations not to exceed 25%) for 15 min at 2.0 m$\ell$ for 1 min.

The ICE separator columns are reconditioned the same as are the cation guard columns. Use 1.0 $M$ HCl to remove mono- and divalent cations. Use 0.1 $M$ sodium tartarate to remove metallic contaminants, but use no more than 5 to 10% acetonitrile to remove organics. For a standard, a mixture of sulfate and formate is satisfactory. Simply remeasure the k' of formate, using the retention time for sulfate to estimate the column void volume.

To recondition MPIC columns, simply wash the guard column with 20 m$\ell$ of 90% acetonitrile or 90% methanol. Re-equilibrate with the desired eluent and remeasure k' for a standard. For example, when using MPIC to determine $Au(CN)_2^-$, the eluent might be 30% acetonitrile with 1 m$M$ tetrapropyl ammonium hydroxide (TPAOH). Use nitrite to estimate the column void volume, and remeasure k' for $Au(CN)_2^-$. The MPIC guard column may be one of the easiest columns to clean since ionic contaminants are not possible, and because it can be washed with such a high concentration (90%) of acetonitrile or methanol.

Other types of contaminants include small particulate material and polymers. They can make it past coarse filters and plug up the top column bed support. This can lead to increased column operating pressures. Fortunately, the bed supports are easily removed and replaced with a new support. In summary, separator columns can be protected by:

1. Using a guard column
2. Filtering the samples before injection
3. Diluting the samples as much as possible

## REFERENCES

1. **Horvath, C.,** Pellicular ion exchangers, in *Bonded Stationary Phases in Chromatography,* Grushka, E., Ed., Ann Arbor Science, Ann Arbor, Mich., 1974.
2. **Moore, S. and Stein, W. H.,** Chromatography of amino acids on sulfonated polystyrene resins, *J. Biol. Chem.,* 192, 663, 1951.
3. EPA Test Method 300.0, The determination of inorganic anions in water by ion chromatography, Environmental Protection Agency, Washington D.C., 1979.
4. **Munger, J. W., Tiller, C., and Hoffmann, M. R.,** Identification of hydroxymethylsulfonate in fog water, *Science,* 231, 247, 1986.
5. **Small, H., Stevens, T. S., and Bauman, W. C.,** Novel ion exchange chromatographic method using conductimetric detection, *Anal. Chem.,* 47, 1801, 1975.
6. **Hirs, C. H. W.,** Amino acid analysis, *Methods Enzymol.,* 47, 3, 1977.
7. **Pohl, C. A. and Johnson, E. L.,** Ion chromatography — the state-of-the-art, *J. Chromatogr. Sci.,* 18, 442, 1980.
8. **Smee, B., Hall, G., and Koop, D.,** Analysis of fluoride, chloride, nitrate, and sulfate in natural waters by ion chromatography, *J. Geochem. Explor.,* 10, 245, 1978.
9. **Muir, A.,** Analysis of inorganic ions by ion chromatography, *Sci. Technol.,* 16, 19, 1978.
10. **Stevens, T., Turkelson, V., and Albe, W.,** Determination of anions in boiler blow-down water by ion chromatography, *Anal. Chem.,* 49, 1176, 1977.
11. **Jansen, K.,** Chromatography: use in the inorganic analysis by ion chromatography, *GIT Fachz. Lab.,* 22, 1062, 1978.
12. **Acciana, T. and Maddalone, R.,** Chemical analysis of wet scrubbers utilizing ion chromatography, EPA Report/600/7-79/151, Environmental Protection Agency, Washington, D.C., 1979.
13. **Otterson, D.,** Application of ion chromatography to the study of hydrolysis of some halogenated hydrocarbons at ambient temperatures, *NASA Tech. Memo,* 1978.
14. **Holcombe, L. and Terry, J.,** Application of ion chromatography for the analysis of lime/limestone based sulfur dioxide scrubber solutions, *Proc. Annu. Meet. Air Pollut. Control Assoc.,* 5, 78, 1978.
15. **Rawa, J. and Henn, E.,** Characterization of industrial process waters and water formed deposits by ion chromatography, *Proc. Int. Water Conf. Eng. Soc. West Pa.,* 40, 213, 1979.
16. **Sawicki, E., Mulik, J. D., and Wittgenstein, E.,** *Ion Chromatographic Analysis of Environmental Pollutants,* Vol. 1 and 2, Ann Arbor Science, Ann Arbor, Mich., 1978, 1979.
17. **Koch, W.,** Complication in the determination of nitrite by ion chromatography, *Anal. Chem.,* 51, 1571, 1979.
18. Environmental Protection Agency, *Quality Assurance Handbook for Precipitation Chemistry Measurement Systems,* EPA 600/4-82-042a + b, Contract No. 68-02-3262, Environmental Monitoring and Support Laboratory, Research Triangle Park, N.C., 1982.
19. **Fitchett, A. W.,** *Analysis of Rain by Ion Chromatography. Sampling and Testing of Rain,* ASTM STP 823, Campbell, S. A., Ed., American Society for Testing and Materials, 1983, 29.
20. **Riviello, J. M., Rocklin, R. D., and Pohl, C. A.,** Gradients in ion exchange chromatography, Paper at 37th Pitt. Conf. Expo. 1986.

21. **Rocklin, R. D., Pohl, C. A., and Schibler, J. A.,** Gradient elution in ion chromatography, 1987 Pittsburgh Conference.
22. **Wheaton, R. W. and Bauman, W. C.,** Ion exclusion chromatography, *Ind. Eng. Chem.,* 45, 228, 1953.
23. **Johansson, I. M., Wahlund, K.-G., and Schill, G.,** Reversed phase ion pair chromatography of drugs and related organic compounds, *J. Chromatogr.,* 149, 281, 1981.
24. **Franklin, G. O.,** Development and application of ion chromatography, *Am. Lab.,* 6, 65, 1985.
25. **Pohlandt, C.,** Determination of cyanides in metallurgical process solutions by ion chromatography, *S. Afr. J. Chem.,* 37, 133, 1984.
26. **Mirna, A., Wagner, H., Klotzer, E., and Fausel, E.,** Zur Anwendung der Ionenchromatographie bei der Untersuchung von Fleisch und Fleischwaren (Application of ion chromatography in the study of meat and meat products), *Lebensmittelchem. Gerichtl. Chem.,* 38, 18, 1984.
27. **Smith, R. E. and Smith, C. H.,** Automated ion chromatography, *LC,* 3, 578, 1985.
28. **Stevens, T. S. and Small, H.,** Surface sulfonation of styrene divinylbenzene optimization of performances in ion chromatography, *J. Liq. Chromatogr.,* 1, 123, 1978.
29. **Hansen, L. C. and Gilbert, T. W.,** Theoretical study of support design for high speed liquid and ion exchange chromatography, *J. Chromatogr. Sci.,* 12, 464, 1974.
30. Dionex, Technical note 17, Anion fiber suppressor-2, 1984.
31. **Stillian, J. R.,** Principles and theory of membrane suppressors, Paper at 37th Pitt. Conf. Symp., 1986.
32. **Haak, K. K.,** The use of ion chromatography in the electroplating industry, *Prod. Finish.,* 37, 3, 13, 1984.
33. **Haak, K. K. and Franklin, G. O.,** Rapid analysis and troubleshooting of gold, copper, and nickel plating bath chemistry by ion chromatography, *Am. Electr. Soc. Conf. Proc.,* 1-6, 1983.
34. **Pacholec, F., Rossi, D. T., Ray, L. D., and Vazopolos, S.,** Characterization of a phosphonate-based sequestering agent by ion chromatography, Liq. Chromatogr., 3, 1068, 1985.
35. Dionex Application note 9, Analysis of organic and inorganic ions in milk products, 1979.
36. **Slingsby, R. and Riviello, J.,** Advances in cation analysis by ion chromatography, Liq. Chromatogr., 1, 354, 1983.
37. **Cheng, K. L., Ueno, K., and Imamura, T.,** *CRC Handbook of Organic Analytical Reagents,* CRC Press, Boca Raton, Fla., 1982.
38. **Martell, A. E. and Smith, R. M.,** *Critical Stability Constants,* Plenum Press, New York, 19, 1977.
39. **Iskandarani, Z. and Pietrzyk, D. J.,** Ion interaction chromatography of organic anions on a poly(styrene-divinylbenzene) adsorbent in the presence of tetraalkonium salts, *Anal. Chem.,* 54, 1065, 1982.
40. **Mohammed, H. Y. and Cantwell, F. F.,** Liquid chromatographic analysis of pharmaceutical syrups using pre-columns and salt adsorption on amberlite XAD-2, *Anal. Chem.,* 50, 491, 1978.
41. **Pietrzyk, D. J., Kroeff, E. P., and Rotsch, T. D.,** Effect of solute ionization on chromatographic retention on porous polystyrene copolymers, *Anal. Chem.,* 50, 497, 1978.
42. **Kroeff, E. P. and Pietrzyk, D. J.,** Investigation on the retention and separation of amino acids, peptides, and derivatives on porous copolymers by high performance liquid chromatography, *Anal. Chem.,* 50, 502, 1978.
43. **Hearn, M. T. W.,** *Ion Pair Chromatography: Theory and Biological and Pharamaceutical Applications,* Marcell Dekker, New York, 1985.
44. **Johnson, E. L. and Haak, K. K.,** Anion analysis by ion chromatography, *Conf. Proc. Liq. Chromatogr. Environ. Anal.,* 263, 1984.
45. **Westerlund, D. and Theodorsen, A.,** Reversed phase ion pair chromatography of naphthaleneacetic acid detivatives with water and an organic modifier as the mobile phase, *J. Chromatogr.,* 144, 27, 1977.
46. **Witmer, D. P.,** Simultaneous analysis of tartarazine and its intermediates by reverse phase liquid chromatography, *Anal. Chem.,* 47, 1422, 1975.
47. **Gloor, R. and Johnson, E.,** Practical aspects of reverse phase ion pair chromatography, *J. Chromatogr. Sci.,* 15, 417, 1977.
48. **Edwards, P. and Haak, K. K.,** A pulsed amperometric detector for ion chromatography, *Am. Lab.,* 4, 78, 1983.
49. **Annable, P. and Smith, R. E.,** Ion chromatography of alkaline soak cleaners, *Plat. Surf. Finish.,* 73, 126, 1986.
50. **Davis, W. R., Smith, R. E., and Yourtee, D.,** Determination of organics in plating solutions using an ion chromatograph, *Metal Finish.,* 63, Sept. 1986.
51. **Tarter, J. G.,** The determination of non-oxidizable species using electrochemical detection in ion chromatography, *J. Liq. Chromatogr.,* 7, 1559, 1984.
52. **Hughes, S. and Johnson, D.,** Amperometric detection of simple carbohydrates at platinum electrodes in alkaline solutions by application of a triple-pulse potential, *Anal. Chim. Acta,* 132, 11, 1981.
53. **Inamuri, T., Tanabe, S., Toida, T., and Kawanishi, I.,** High performance liquid chromatography of inorganic anions using Fe(III) as a detection reagent, *J. Chromatogr.,* 250, 55, 1982.

54. **Yamagishi, M. and Yoshida, T.,** Ninhydrin reaction. I. Colorimetric determination of alpha amino acids, *J. Pharm. Soc. Jpn.,* 3, 65, 1953.

55. **Roth, M.,** Fluorescence reaction for amino acids, *Anal. Chem.,* 43, 880, 1971.

56. **Jandera, P. and Churacek, K. J.,** *Gradient Elution in Column Liquid Chromatography,* Elsevier, New York, 1985, 6.

57. **van Deemter, J. J., Zuiderweg, E. J., and Klinkenberg, A.,** *Chem. Eng. Sci.,* 5, 21, 1957.

58. **Snyder, L. R. and Kirkland, J. J.,** *Introduction to Modern Liquid Chromatography,* 2nd ed., John Wiley & Sons, New York, 1979, 30.

59. **Nickerson, R. F.,** The combining ratio of boric acid and alkali borate with mannitol., *J. Inorg. Nucl. Chem.,* 30, 144, 1968.

60. **Doury-Berthod, M., Giampaoli, P., Pitsch, H., Sella, C., and Poitrenaud, C.,** Theoretical approach of dual column ion chromatography, *Anal. Chem.,* 5, 225, 1985.

Chapter 2

# CONVENTIONAL METHODS

## I. CONDUCTIVITY DETECTION

After ion chromatography was first introduced, it rapidly became known as a technique that used ion exchange separator columns and suppressor columns. The introduction of suppressor columns was especially noteworthy, because it enabled the use of conductivity detectors. The conductivity detector became an almost universal detector for ions. It enabled the simultaneous separation and detection of at least seven common anions. Environmental scientists were able to simultaneously determine chloride, phosphate, nitrate, and sulfate in water samples. Electroplaters were able to determine chloride in concentrated acids and bases. Biochemists could determine organic anions in physiological fluids. The purpose of this chapter, therefore, is to provide information about this, the oldest and most well-known form of ion chromatography.

Even after the introduction of alternative separation and detection techniques, the vast majority of the publications in the field of ion chromatography described the use of ion exchange separator columns, and conductivity detection. American Society of Testing and Materials (ASTM) and Environmental Protection Agency (EPA) standard test methods, incorporate ion exchange with chemically suppressed conductivity detection. To this day, there are many publications appearing that still utilize the "conventional" separation technique of ion exchange together with chemically suppressed conductivity detection. Many of the more recent applications, however, use different separator and suppressor columns. To the biochemist or chemical engineer just beginning to become familiar with ion chromatography, it must be confusing to read one paper about an AS-1 column, another about a "standard" separator and suppressor column, and still others about AS-2, AS-3, AS-4, or AS-5 separators. The confusion must be even worse when terms such as "packed-bed suppressor", "fiber suppressor", or "micromembrane suppressor" are used. It can make one believe that the field of ion chromatography is complicated, which it is not. Biochemists who have used DEAE-cellulose or CM-cellulose to purify enzymes will discover that no new concepts are needed to understand ion chromatography. Chemical engineers who use water softeners or zeolite, will also find themselves quite familiar with the basic principles of ion exchange, needed to understand chemically suppressed ion chromatography.

The reason for the apparent complexity is simply that new hardware is being developed so rapidly that very many new terms are coined. Inventors of novel separation and detection methods are required to create a new vocabulary to obtain a patentable product with a distinct trademark. The purpose of this chapter, then, is to explain some of this new vocabulary. In so doing, this chapter can become a reference to both present and potential users of ion chromatographs. By developing an understanding of the different separator columns, the analyst can better understand the ion chromatographic literature. In many cases, any of several separator columns can be used to achieve acceptable results. It may be that one analyst simply used the columns that were available at the time. It is important to understand the different properties of the columns when trying to reproduce previously reported separations. It is often quite possible that the same separations can be performed faster and more efficiently by using newer hardware. On the other hand, it is important to know the capabilities and limitations of the older hardware when critically evaluating the significance of previous work. Quite often, separations can be performed faster and easier using more modern columns. When analyzing particularly difficult samples, column lifetime can be enhanced significantly by using the best column. For example, the AS-4A was especially designed for analyzing soil samples that could poison the AS-4 column. By understanding the differences between the various separators, the analyst will be able to more cleverly develop important new analytical methods.

## II. ANALYSIS OF COMMON ANIONS

The first anion separator marketed, the AS-1, was based on the anion exchange column developed by Stevens and Small.[1] In the first ion chromatography papers, this column was referred to as the "standard" separator. The AS-1, or standard anion separator column, rapidly became the workhorse for ion chromatography. There are more publications that use the AS-1 than any other separator column. This, unfortunately, may lead one to believe that the AS-1 is still the separator column of choice for some analyses; this is not the case. The AS-1 has been replaced. It was first replaced by the AS-3, and then by the AS-4 and AS-4A. However, this does not lessen the validity of the analytical results obtained using an AS-1. It simply means that anions can be separated faster and more efficiently with the more modern anion separator columns. It also means that the analyst should be familiar with all the anion separator columns. In this way, the previous literature can be understood and new analytical methods can be developed using the optimum columns.

To achieve an understanding of the different separator columns it is useful to know their physical and chemical composition. The AS-1 has a polystyrene/divinylbenzene base resin core particle with a 25-$\mu$m diameter with 5% cross linking. The R group in the $-NR_3$ is classed as hydrophobic[2] and the latex beads which are attached to them are of medium size. In the original work by Stevens and Small, NaOH was used as the eluent for anion separations. Potentially, NaOH could be the ideal eluent for anions because it is converted to water by the suppressor column. There was one problem, however, the hydroxide ion does not have a strong affinity for anion exchange sites on the anion separator columns. As a result, relatively high concentrations of NaOH are required to elute divalent anions such as sulfate and oxalate. Using the original packed-bed suppressor columns, this meant that the suppressors had to be regenerated quite often. To overcome this difficulty, a mixture of sodium bicarbonate (3 m$M$) with sodium carbonate (2.4 m$M$) soon became popular. Because this concentration is much lower than that required when NaOH is used as the eluent, the packed-bed suppressors did not require such frequent regeneration. Even after the packed-bed suppressors were replaced by fiber suppressors, the bicarbonate/carbonate eluent remained quite popular. The suppressor column converts sodium bicarbonate and carbonate to carbonic acid. Even though carbonic acid has higher conductivity than the water produced when NaOH eluent is used, the background conductivity is still low enough to provide sub-ppm detection limits for most anions. The fiber suppressors simply do not have a high enough capacity to suppress the high concentration of NaOH that would be needed to elute divalent anions in a reasonable time. When NaOH was first used as the eluent, peaks due to the analyte anions eluted late and had poor peak shapes, compared to what was obtainable with the bicarbonate/carbonate eluent. The analytes could be eluted with fewer microequivalents of carbonate than hydroxide. As a result, the vast majority of publications that describe anion analysis by ion chromatography use a mixture of sodium bicarbonate and sodium carbonate as eluent.

Using a bicarbonate/carbonate eluent and the AS-1 column, the seven anions, fluoride, chloride, nitrite, phosphate, bromide, nitrate, and sulfate, can be effectively separated in 16 min as shown in Figure 1. These seven anions have been used as standards so often in anion chromatography, that they have become known as the seven common anions. If one surveys the ion chromatography literature or attends scientific conferences, they are led to believe that this separation and detection of seven common anions is the essence of ion

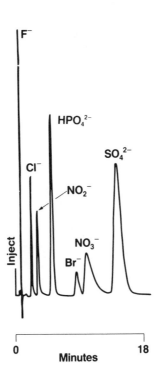

FIGURE 1. Standard Anion Analysis. Column: AS-1, Detection: Chemically Suppressed Conductivity.

chromatography. The majority of all papers published in the field of ion chromatography describe the separation and detection of at least two of the "seven common anions". This is partly because the development of a method to simultaneously separate and detect fluoride, chloride, nitrite, phosphate, bromide, nitrate, and sulfate in one run, was extremely significant. It is quite likely that even today, most ion chromatographs are being used, at least part of the time, to separate and detect some of the seven common anions. Chloride is perhaps one of the most important. When a new hand-held calculator is first turned on, any chloride ions that are present in its microcircuit will migrate. If present at too high a concentration, they can cause short circuits, and the calculator will malfunction. As a result, manufacturers of hand-held calculators continuously analyze the plating solutions for chloride contaminants. When environmental scientists want to determine the extent of acid rain, they analyze water samples for chloride. When toxicologists analyze physiological fluids using ion chromatography, there is always a large chloride peak. Thus, it is important to reemphasize that when most people think of ion chromatography, they think of analyzing samples for common everyday anions. To the inorganic chemist this may mean chloride, nitrate, phosphate, and/or sulfate. Certainly ion chromatography can perform these determinations admirably. This was evidenced in a U.S. Geological Survey. In this interlaboratory testing program, fluoride, nitrate, and sulfate were determined using other analytical methods (colorimetry, ion-selective electrodes, turbimetric, and gravimetric). The results are summarized in Tables 1 to 3. The relative standard deviations ranged from 3.4 to 12.4% (for fluoride). This test was performed using the AS-1 separator and packed-bed suppressor. Since then, newer columns have been introduced which enable much better precision, but much of the ion chromatographic literature describes the use of just these columns; much significant scientific knowledge was achieved using these columns. It is almost by accident, that ion chromatography was first put to practical application by inorganic chemists. In fact, there are other anions which are every bit as common. There are other anions such as creatine

**Table 1**
**U.S. GEOLOGICAL SURVEY**
**INTERLABORATORY TESTING PROGRAM**

| Determination: $F^-$ (Method) | Mean | SD | N |
|---|---|---|---|
| Colorimetric, SPADNS | 2.12 | 0.32 | 4 |
| Ion Chromatography | 2.10 | 0.26 | 3 |
| Ion Selective Electrode, Automated | 1.95 | 0.14 | 10 |
| Ion Selective Electrode, Manual | 1.96 | 0.19 | 36 |
| **Overall average** | **1.99** | **0.21** | **62** |

*Note:* SD, standard deviation; N, number.

**Table 2**
**U.S. GEOLOGICAL SURVEY INTERLABORATORY**
**TESTING PROGRAM**

| Determination: $NO_3^-$ (Method) | Mean | SD | N |
|---|---|---|---|
| Colorimetric, Brucine | 3.943 | 0.407 | 17 |
| Colorimetric, Cadmium Reduction, Diazotization | 4.020 | 0.377 | 40 |
| Colorimetric, Hydrazine Reduction, Diazotization | 3.717 | 0.411 | 4 |
| Ion Chromatography | 3.972 | 0.336 | 4 |
| **Overall Average** | **3.975** | **0.392** | **68** |

*Note:* SD, standard deviation; N, number.

**Table 3**
**U.S. GEOLOGICAL SURVEY INTERLABORATORY**
**TESTING PROGRAM**

| Determination: $SO_4^{2-}$ (Method) | Mean | SD | N |
|---|---|---|---|
| Colorimetric, Methyl Thymol Blue, Automated | 217.5 | 8.8 | 20 |
| Gravimetric, Barium Sulfate | 225.2 | 13.7 | 12 |
| Ion Chromatography | 227.2 | 7.8 | 6 |
| Thorin Titration | 220.0 | 0.0 | 3 |
| Turbidimetric, Barium Sulfate | 222.2 | 15.2 | 26 |
| **Overall average** | **221.9** | **12.7** | **68** |

*Note:* SD, standard deviation; N, number.

phosphate, ribulose-1,5-diphosphate, glucose-6-phosphate, and even uridine mono-, di-, and triphosphate that are present in ample abundance throughout the biosphere.

Certainly, then, there are many other anions which can also be separated on the AS-1 and be detected by chemically suppressed conductivity. A partial list of such anions is given in Table 4. The anions are listed in their approximate elution order. Using a 3-m$M$ sodium bicarbonate/2.4-m$M$ sodium carbonate eluent, many of these anions would co-elute. For example, if the first 6 anions listed (fluoride through chlorite) were present in the same sample, and were injected on the AS-1, only one peak would be seen. The seventh entry in Table 4, organophosphates, requires further elaboration.

The term "organophosphates" is, of course, a generalization. In fact, no attempt has yet been made to separate the vast majority of organophosphates. Because so many phosphorylated compounds have profound biological significance, and attempts to determine them by ion chromatography will be made in the near future, it is important to review the chromatographic properties of those that have already been investigated.

**Table 4**
**ANIONS DETERMINED BY**
**ION CHROMATOGRAPHY**
**WITH CONDUCTIVITY**
**DETECTION**

| | |
|---|---|
| Fluoride | Malonate |
| Formate | Chlorate |
| Acetate | Nitrate |
| Propionate | Maleate |
| Iodate | Itaconate |
| Chlorite | Tartarate |
| Organophosphates | Sulfate |
| Hypophosphite | Sulfite |
| Chloroacetate | Dioxytartarate |
| Bromate | Ascorbate |
| Chloride | Trichloroacetate |
| Glycolate | Fumarate |
| Pyruvate | Arsenate |
| Nitrite | Oxalate |
| Dichloroacetate | Fluoroborate |
| Phosphite | Selenate |
| Phosphate | Thiosulfate |
| Selenite | Tungstate |
| Succinate | Molybdate |
| Bromide | Chromate |

To begin with, Schiff et al.[3] used the AS-1 column to separate the nerve-agent-related phosphonic acids, isopropyl methyl phosphonic acid (IMPA), ethyl methyl phosphonic acid (EMPA), and methyl phosphonic acid (MPA). Using the standard 3-m$M$ sodium bicarbonate/2.4-m$M$ sodium carbonate eluent, fluoride co-elutes with IMPA and EMPA. Also, chloride co-elutes with MPA. Because all these ions are weakly retained on the AS-1, a weaker eluent could be used. It was found that fluoride could be separated from IMPA and EMPA using a 5-m$M$ sodium tetraborate eluent; IMPA and EMPA both elute before fluoride. It was also found that chloride could be separated from EMPA and MPA. Using the standard 3-m$M$ sodium bicarbonate/2.4-m$M$ sodium carbonate eluent, fluoride co-elutes with IMPA and EMPA. Also, chloride co-elutes with MPA. Because all these ions are weakly retained on the AS-1, a weaker eluent could be used. It was found that fluoride could be separated from IMPA and EMPA using a 5-m$M$ sodium tetraborate eluent. IMPA and EMPA both elute before fluoride. It was also found that chloride could be separated from MPA by using a 2.4-m$M$ sodium bicarbonate/1.8-m$M$ sodium carbonate eluent. MPA was found to elute just after chloride. For surface and ground water analysis, however it was found that a 10 m$M$ NaOH eluent was preferred, because the resolution between MPA and chloride was even better than when the 2.4-m$M$ sodium bicarbonate/1.8-m$M$ sodium carbonate eluent was used. The point here, is that when separating anions that are weakly retained, a weaker eluent is required. Three possible solutions exist. First, use more dilute sodium bicarbonate/sodium carbonate, second, use sodium tetraborate, or finally, NaOH could be used as the eluent.

The alkyl phosphonic acids are not the only organic phosphates that have been shown to be weakly retained on the AS-1. Lash and Hill[5] found that dibutylphosphoric acid (DBP) and tributylphosphoric acid (TBP) are also weakly retained. These compounds are found in the nuclear fuel reprocessing industry. These workers used 3 m$M$ sodium carbonate/1 m$M$ NaOH as the eluent. This high pH eluent fully ionizes phosphate in their samples to $PO_4^{3-}$. As a result, phosphate is more strongly retained and is well separated from the DBP and TBP.

In addition, several phosphorylated compounds of biological significance have been shown to elute early off the AS-1.[4] Creatine phosphate, glycerol-3-phosphate, and phosphoenol pyruvate were found to elute between fluoride and chloride. Glucose-1-phosphate co-elutes with chloride, but glucose-6-phosphate elutes shortly afterwards, followed by fructose-6-phosphate and fructose-1,6-diphosphate. There are many more biologically significant phosphorylated compounds which have either been separated on other anion separators, or have simply never been investigated by ion chromatography.

One of the more extensive listings of anions that can be determined was reported by Deister and Runge[6] in their description of applications in the iron and steel industry. They described the determination of fluoride, formate, acetate, iodate, chlorate, chloride, bromate, nitrite, phosphate, bromide, nitrate, sulfite, sulfate, and oxalate. These anions eluted in approximately that order. They co-elute using 3.0 m$M$ sodium bicarbonate/2.4 m$M$ sodium carbonate. To separate the more weakly retained anions, a weaker eluent is required. Using 1.5 m$M$ sodium bicarbonate, they were able to separate fluoride, iodate, chlorate, formate, bromate, chloride, and nitrite. Using this weaker eluent, chloride elutes in 22 min and nitrite in 30.7 min. Using the stronger bicarbonate/carbonate eluent, chloride elutes in just under 2 min, and divalent anions can still be eluted. Oxalate elutes in 31.2 min. Chloride, nitrate, and sulfate were determined in potable water and wastewater. Chloride, phosphate, nitrate, and sulfate were determined in cooling water and steam condensate. Sulfite was determined in waste gas. Thus, ion chromatography was demonstrated to be a very useful analytical tool in the steel industry.

Sulfite is also used in the food and pharmaceutical industries as an antioxidant. Lindgren et al.[7] demonstrated that sulfite can be stabilized in samples by adding formaldehyde. By using a 1:1 mole ratio of formaldehyde to sulfite, oxidation from atmospheric oxygen can be prevented. In addition, formaldehyde forms a complex with sulfite. This complex is better resolved from sulfate than is the free sulfite. The HCHO-SO complex eluted at 10.3 min, whereas sulfate eluted at 14 min.

Another anion separated on the AS-1 was sulfamate. Sulfamate is a flame retardant, a weed killer, and is sometimes used as a nickel salt in electroplating. The OSHA standard for an 8-hr exposure time of sulfamate is 15 mg/m[3]. Using 3 m$M$ sodium carbonate/2.0 m$M$ sodium bicarbonate eluent flowing at 2.3 m$\ell$/min, Bodek and Smith[8] found that sulfamate elutes between fluoride and chloride. They obtained a linear calibration plot from 5 to 100 ppm sulfamate.

Tungstate and molybdate have also been determined in water samples using an AS-1. Ficklin first preconcentrated the molybdate and tungstate on a Chelex® 100 column.[9] They added 5 m$\ell$ of concentrated nitric acid to 1.0 m$\ell$ of a water sample, then passed it through 2 g of Chelex®. The tungsten and molybdenum were eluted off the Chelex® with 7 m$\ell$ of concentrated ammonium hydroxide and 10 m$\ell$ of deionized water. After boiling-off the ammonia, the sample could then be injected on the ion chromatograph. Tungstate and molybdate were separated on an AS-1. Using a 6-m$M$ sodium carbonate eluent flowing at 1.5 m$\ell$/min, tungstate eluted in approximately 9 min, and molybdate in approximately 12 min.

Selenium can also be determined in aqueous solution as selinite and selenate. Chakraborti et al.[10] used the AS-1 with 8 m$M$ sodium carbonate flowing at only 0.46 m$\ell$/min. Selenite eluted in 12 min, and selenate in 21 min.

The AS-1 column demonstrates a point made earlier. It is a prime example of a piece of ion chromatography hardware that is mentioned often in the chemical literature (especially in the 1970s and early 1980s), but which has since been supplanted by the AS-3, 4, and 4A columns. The comparatively large size of the polystyrene/divinylbenzene core particle (25 μm) kept the back-pressure of the ion chromatograph low enough (below 800 psi) that low-pressure fittings and a less precise pump could be used. From 1976 to 1980, this

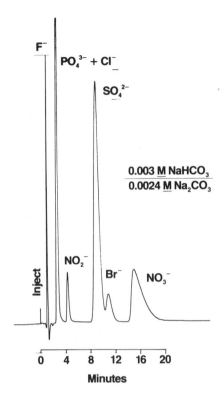

FIGURE 2.   Anion Standard — S-2 Column. Column: AS-2. Note the different elution order compared to the AS-1.

hardware, in the form of the Models 10, 12, 14, and 16 ion chromatographs, was state-of-the-art.[11] Laboratories trying to repeat the work reported in this time-period, will find that ion chromatography hardware has changed significantly. The higher-efficiency columns and pumps, along with higher-pressure fittings, enable the same separations that required 16 min to be done in 7 min.

The second anion separator marketed by Dionex was the AS-2. This column was first called the "brine column" because it was especially useful in determining nitrate in brine samples. Although the AS-2 is still quite useful in analyzing brines, it has found few other applications. The AS-2 has the same 25-$\mu$m polystyrene/divinylbenzene core particle size as the AS-1, but the R group on the $-NR_3$ is more hydrophobic. The elution order of the seven common anions with a bicarbonate/carbonate eluent is different, as seen in Figure 2. Both bromide and nitrate elute after sulfate. For brine analysis, this meant that the nitrate peak is well separated from the large chloride peak from the brine. In addition to nitrate, phosphate and sulfate are separated well enough from chloride to permit their determination in brine samples.[12] To determine phosphate, it is necessary to use a higher pH eluent. If the sodium bicarbonate/sodium carbonate eluent was used, the phosphate would exist as $HPO_4^{2-}$ and would elute too close to the $Cl^-$ in the brine to be accurately determined. By using a higher pH eluent, 5 m$M$ sodium carbonate/4 m$M$ sodium hydroxide, the phosphate exists as the trivalent $PO_4^{3-}$ and is more strongly retained on the AS-2 column (referred to as the S-2 resin in Reference 11). The retention time or capacity factor for $Cl^-$ is almost unchanged using this higher pH eluent, so that the phosphate peak is much better separated and can be easily quantitated. Because so few articles have appeared that use the AS-2, and because of the development of more efficient, modern columns, it is likely that the AS-2 will never see widespread use. It is important to realize, however, that it can still be the

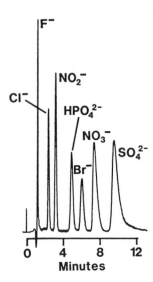

FIGURE 3.    Anion Standard on the AS-3. Note the increased column efficiency and faster analysis time.

column of choice when a unique selectivity is required. This is especially true when analyzing brines.

The third anion separator column, the AS-3, was developed as an improvement on the AS-1. The AS-3 provides faster run-times and greater efficiency for separating anions. In fact, the AS-3 was originally called the "fast-run" column. The AS-3 still has the 25-$\mu$m polystytene/divinylbenzene core particle size, but it is only 2% cross-linked and the latex particle size is smaller than with the AS-1 or 2. The specificity of the AS-3 very closely resembles that of the AS-1. In fact, the AS-3 was originally called the fast run column because it separated the seven common anions much faster than the AS-1. As shown in Figure 3, the chromatogram of the seven common anions looks the same as with the AS-1 except that the anions elute faster (compare to Figure 1). Not only do the anions elute faster, but the column efficiency is superior to the AS-1. The resolution and height equivalent per theoretical plate, is improved. The AS-3 produces no more back-pressure than the AS-1 or 2 (because they all have the same 25-$\mu$m core particle size). This permitted the use of the same pump and lower-pressure fittings of the Models 10, 12, 14, and 16.

The AS-3 column has also been used in the most current ion chromatograph (the Series 4000) and a gradient elution. It was used to separate nuceloside mono-, di-, and triphosphates as shown in Figure 4. The nucleotides are retained rather strongly on the AS-3, so they require a relatively strong eluent. This is why 0.2 $M$ potassium phosphate was used. The phosphate eluent is, of course, transparent to the UV detector used in this analysis. There is no need for a suppressor column when using the UV detector.

Tan and Dutrizac[13] used the AS-3 to determine arsenic (III) and arsenic (V) in ferric chloride leaching media. Ferric chloride is used to leach base metal sulfide concentrates in smelting operations. They used a 3.5-m$M$ sodium carbonate/1.0-m$M$ sodium hydroxide eluent flowing at 2.7 m$\ell$/min. With this eluent, nitrate elutes at about 5.9 min and arsenate at 12.4 min. The metals in the sample were removed prior to injection on the ion chromatograph. This was done by passing the leachate sample through Dowex® 50-X4 cation exchange resin. First, the amount of arsenate in the sample was determined. Then, the arsenite in the sample was oxidized to arsenate by adding aqua regia. The amount of arsenite present in the original sample was calculated by difference, i.e., the amount of arsenate found in the first analysis was subtracted from the amount of total arsenate found after oxidizing the arsenite.

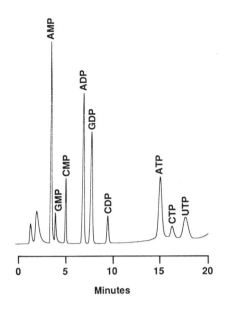

Column: HPIC-AS3

Detector: UV, 254nm

Gradient: 5 to 95% B in 24 min.

A: Water

B: 0.2M $K_2HPO_4$, pH 4.5

Flow rate: 2mL/min.

FIGURE 4.    Separation of Nucleotides. Column: AS-3, Detection: UV, 254 nm. Gradient elution, start with deionized water, increase to 0.2 $M$ potassium phosphate, pH 4.5.

Hoover and Yager[14] used the AS-3 column in a different way to determine As(V) in tap water. They injected the sample on an AS-3 column, but the peak due to As(V) (i.e., arsenate) overlapped with the peak due to sulfate. As the peak due to sulfate and arsenate appeared on the chromatogram, a valve was switched so that the effluent off the AS-3 was sent to an AG-3, i.e., a guard column. (The AG-3 is simply a short AS-3 column.) By collecting the portion of effluent containing arsenate and sulfate, it became possible to perform a second separation. The arsenate and sulfate which would not separate under the chromatographic conditions used on System 1, which could then be separated under different chromatographic conditions on System 2. This is a form of two-dimensional chromatography. This particular form, in which a portion of a chromatogram is collected and reinjected on a similar column, is called "heart-cutting". Hoover and Yager expanded the original work by also determining selenite and arsenate in groundwaters.[15] The National Interim Primary Drinking Water Regulations mandate a maximum total arsenic at 50 ppb and Selenium at 10 ppb. These anions, as well as chloride, nitrate, and sulfate, were separated by heart-cutting using an AG-1 guard column, the AS-3 separator column, and the original packed-bed suppressor.

Selenate elutes after sulfate with the standard bicarbonate/carbonate eluent. Selenite elutes between chloride and nitrate. Arsenate can be made to elute after sulfate by using the alkaline eluent, 3.5 m$M$ sodium carbonate/2.6 m$M$ sodium hydroxide. Eluent conditions had to be optimized for each anion.

The AS-3 was also used by Phillipy and Johnston[16] to determine phytic acid, a hexa-phosphocyclohexane. They assayed phytic acid in infant formula powder, soy flour, soy isolate, wheat bran, and wheat bread. After homogenizing the sample, they extracted it with 1.2% HCl. After dilution, the sample was injected on an AS-3 that had been pre-equilibrated with 0.11 $M$ nitric acid eluent. The phytic acid was detected by post-column reaction with ferric chloride/perchloric acid. The basis for the detection was the formation of a colored complex between the ferric ion and the phytic acid. The phytic acid eluted in about 7 min under these conditions.

Almost the time that the AS-3 was introduced, the first anion fiber suppressors also appeared. This eliminated the need for off-line regeneration required of packed-bed sup-

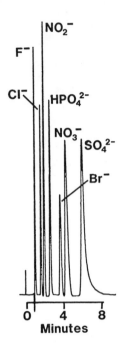

FIGURE 5.   Anion Standard on the AS-4. Note increased column efficiency.

pressors. The fiber suppressor was continually regenerated with the flow of regenerant being gravity-fed by a large reservoir of sulfuric acid above the ion chromatograph. A two-way valve was used to control the flow of regenerant. The continuously regenerating fiber suppressors, together with the faster runs attainable from the AS-3, opened the door to automation. All that was needed was to install a microprocessor to control the pump and valve that turned regenerant flow, off and on. This first level of automation became known as the Auto-Ion System 12 analyzer. All the analyst was required to do was dilute, filter, and load the samples into an autosampler. After this was done, completely unattended analysis was provided.

The fourth anion separator, the AS-4, helped usher in the Series 2000 ion chromatographs. Unlike the AS-1, 2, and 3 columns, the AS-4 produced higher back-pressure because it had a smaller polystyrene/divinylbenzene core particle size of only 15 μm. This required the introduction of a more precise analytical pump and higher-pressure fittings.

The AS-4 is 3.5% cross linked. The latex particles in the AS-4 are small, similar to the AS-3. The R groups in the $-NR_3$ are hydrophobic. The AS-4 column was introduced together with a whole new series of ion chromatographs called the Series 2000, and the term "high-performance ion chromatography" was coined. The Series 2000 has a much more precise pump and higher-pressure fittings. This was required because of the smaller (15 μm) particle size of the core particle in the AS-4, which produced higher back-pressure (1100 psi) at normal (about 2 mℓ/min) flow-rates. The AS-4 is an even "faster run" column than the AS-3, i.e., all seven common anions elute in the same order but in less time, as shown in Figure 5. This faster analysis time, together with the analytical pump and the elevated operating pressures, were used to justify the introduction of the new term "high-performance ion chromatography".

By now, organic anions, aliphatic and aromatic, began to be separated. The nucleotides cytidine monophosphate (CMP), uridine monophosphate (UMP), adenosine monophosphate (AMP), and guanidine monophosphate (GMP) can be separated on an AS-4 using 2.5 m$M$ sodium phosphate, pH 3.35, adjusted with phosphoric acid as shown in Figure 6. In addition,

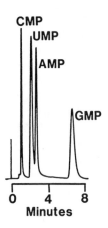

FIGURE 6.  Separation of Nucleoside Monophosphates. Column: AS-4, Detector: UV, 254 nm.

the nucleosides uridine, guanosine, along with CMP and AMP, were separated using 2.9 m$M$ sodium bicarbonate plus 2.3 m$M$ sodium carbonate as the eluent.[2] Although nonionic aromatic compounds are strongly retained on the AS-1, benzoic acid elutes near chloride and nitrite. Four substituted benzoic acids, $o$-methyl hippuric acid, mandelic acid, hippuric acid, and $p$-methyl hippuric acid were separated using two AS-4 columns in series. The eluent used was 5 m$M$ Na$_2$CO$_3$ plus 5 m$M$ NaOH. Chemically suppressed conductivity was used for detection.[2] The elution order was shown to be directly related to the pKa values of these benzoic acids. It should be noted, however, that other benzoic acids, such as $p$-chloro benzoic acid, $m$-bromo benzoic acid, $m$-fluoro benzoic acid, and salicyclic acid are all strongly retained on the AS-4, presumably due to nonionic interactions with the polystyrene/divinylbenzene core particle. Aliphatic carboxylic acids were also separated on two AS-4 columns in series. Formate, succinate, malonate, maleate, tartarate, and oxalate were eluted, in that order, using 2.8 m$M$ sodium bicarbonate/2.2 m$M$ sodium carbonate eluent. These two applications illustrate an important point. Resolution is directly related to the number of theoretical plates. If the number of plates is doubled by using two columns connected to each other, the resolution will improve. This is useful if determining anions which produce overlapping peaks, using one column. A particularly good application of this is the determination of a trace anion in the presence of a large excess of another anion. For example, if trying to determine trace levels of sulfate in reagent grade tartaric acid, the sulfate appears as a small peak or shoulder on the much larger tartarate peak. It is often necessary to inject rather concentrated samples of tartaric acid to keep from diluting the trace sulfate so much that it cannot be detected. The AS-4 column can easily become overloaded, causing irreproducible results. By using two AS-4 columns, not only does the resolution between tartarate and sulfate almost double, but it also takes twice as much tartarate to overload the dual column system. The point to be made here is that the ion chromatograph can handle the additional back-pressure from two separators, and can be so utilized.

The AS-4 column was used by Cox and Tanaka[17] to determine anionic impurities in several concentrated electrolytes. To analyze 50% NaOH, the sample was first passed through a cation exchange resin in the proton form. The Na$^+$ was exchanged for H$^+$, converting the highly ionic NaOH to the nonionic H$_2$O. This is necessary because the AS-4 has limited anion exchange capacity. The AS-4 would be overloaded by NaOH solutions as dilute as 1 $N$ or 4%.[18] By converting the NaOH to water, the analyst obtains a dilute solution of anions, which is ideal for ion chromatography. The anions could then be separated on an AS-4 column. To analyze sodium bicarbonate or sodium carbonate, Cox and Tanaka[17] passed the 0.1- to 0.5-*M* solution through a dual ion exchange apparatus consisting of 2 m of Nafion® 811 tubular cation-exchange membrane, and a slurry of Dowex® 50WX4 100/200 mesh cation-exchange resin in the proton form. This ion-exchange method for sample treatment was described in more detail in a subsequent paper. The bicarbonate and carbonate are converted to $CO_2$ and water, so that microgram quantities of chloride, nitrate, and sulfate per kilogram of sodium bicarbonate or carbonate can be determined. Cox and Tanaka also determined sulfate in 0.1 *M* HCl and in 0.1 *M* NaCl.

The AS-4 was also used by Muller[19] to analyze precipitation samples. Not only the seven common anions, but also formate and acetate were reported to be separated on the AS-4.

Each of the first four anion separator columns have one limitation in common, i.e., a number of large, hydrophobic anions which are strongly retained and cannot be easily determined. The recommended procedure in such cases was to omit the separator column and use only a guard column and a strong eluent. Since a guard column is only a short separator column, the large hydrophobic chromate can be eluted using the relatively strong 6-m*M* sodium carbonate eluent.[20] Similarly, thiocyanate, thiosulfate, and iodide were eluted off an AG-2 (the guard column for the AS-2). Because of the limited number of theoretical plates in such a small column, the seven common anions cannot be separated, but instead, four of them, fluoride, chloride, phosphate, and sulfate, are separated.

Thus, it was a significant advance when the AS-5 column was introduced for the determination of these large, hydrophobic anions. The AS-5 has a 15-μm polystyrene/divinylbenzene core particle that is only 1% cross-linked. The latex beads are small (similar to the AS-3 and AS-4), but the R groups in the $-NR_3$ are hydrophilic. Using the AS-5 and the conventional bicarbonate/carbonate eluent, chromate elutes in about 20 min.[21] This is convenient for automation. The AS-5 and AS-4 columns can be connected to one valve, and a microprocessor can automatically select the desired column. The analyst need only enter on the microprocessor the number of samples to be analyzed on each column. By using the proper programs, each column can be calibrated, with standards and calculations made on all samples. The computer can switch files at the appropriate times to accomodate the calculations for different samples. Thus, it would be possible to run standards on an AS-4 column, then also to have samples injected. The computer will have already used the calibration data to calculate the concentrations of anions in the sample. After analyzing all the samples on the AS-4, the computer can then switch the appropriate valve to select the AS-5. After equilibrating the AS-5 (with the same eluent used for the AS-4), more standards can be injected, this time on the AS-5. The computer will be able to switch files at the appropriate times and automatically make all necessary calculations.

Although the bicarbonate/carbonate eluent can be used with the AS-5 column, peak tailing in the more strongly retained anions can be significant. This is due to hydrophobic interactions between the large hydrophobic anions and the AS-5 column. To minimize these hydrophobic interactions and eliminate peak tailing, 0.75 m*M* *p*-cyanophenol is added to the eluent. As shown in Figure 7, not only does the peak tailing for iodide and thiocyanate decrease, but also they are not so strongly retained on the AS-5. By obtaining taller, sharper peaks, the detection limits are lowered for these and similar anions.

In addition to chromate, other oxy-metal anions can be separated on the AS-5 using the conventional bicarbonate/carbonate eluent as shown in Figure 8. This analysis was performed using an anion fiber suppressor. After the introduction of the micromembrane suppressor, the stronger 10 m*M* NaOH plus 5 m*M* sodium carbonate eluent, could be used. The anion micromembrane suppressor can easily suppress this higher anion flux. As a result, the same four oxy-metal anions can be separated in 11 min, as shown in Figure 9. Because the peaks are larger and sharper, lower detection limits are attainable. In general, most anions are not retained as strongly on the AS-5 as on an AS-4 or 4A. One exception is sugar phosphates.

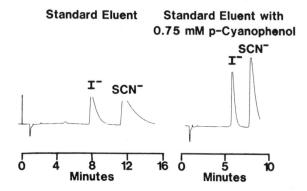

FIGURE 7. Effect of *p*-Cyanophenol on the Chromatography of Iodide and Thiocyanate. Cyanophenol reduces nonionic interactions between the analyte ions and the polystyrene/divinylbenzene core on the AS-5 resin.

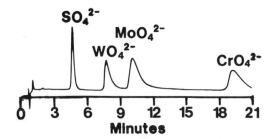

FIGURE 8. Analysis of oxy-metal anions with the AS-5 column. Because the anion micromembrane suppressor (AMMS) was not used (instead a fiber suppressor was used), the eluant concentration could not be increased sufficiently to elute chromate in a reasonable time.

Column: HPIC-CS5
Eluant: 10mM NaOH
       5mM $Na_2CO_3$
Detector: Conductivity/AMMS

| Concentrations | ppm |
|---|---|
| 1. Sulfate ($SO_4^{2-}$) | 10 |
| 2. Tungstate ($WO_4^{2-}$) | 50 |
| 3. Molybdate ($MoO_4^{2-}$) | 50 |
| 4. Chromate ($CrO_4^{2-}$) | 50 |

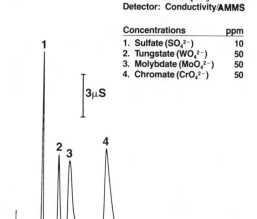

FIGURE 9. Oxy-metal complexes on the same AS-5. Because the AMMS was used, higher eluent concentrations were possible, and chromate elutes in 1 min.

Although not extensively investigated, at least inositol-2-phosphate and glucose-6-phosphate are more strongly retained and are well separated from phosphate on the AS-5 column.[21]

The AS-5 column has also been used to determine gold, palladium, and platinum at the ppb level.[22] Gold, lead, and platinum form anionic complexes with chloride, i.e., $AuCl_4^{2-}$, $PdCl_6^{2-}$, and $PtCl_4^{2-}$. The chloride complexes were formed by digesting the metal sample in hot aqua regia (3:1 $HCl/HNO_3$). The eluent consisted of 0.30 $M$ sodium perchlorate with 0.05 $M$ HCl. $PtCl_4^{2-}$ and $PdCl_6^{2-}$ co-elute at about 4 min, but $AuCl_4^{-}$ elutes at about 10 min. The very strong perchlorate eluent is required because the metal-chloride complexes are so large and hydrophobic that they are strongly retained on the AS-5. To attain parts per billion sensitivity, the sample is first loaded onto a concentrator column using the much weaker 0.2 $M$ HCl eluent. The metal chlorides were then eluted off the concentrator column with the perchlorate eluent. The metal chlorides were detected with a UV detector set at 215 nm. They can also be detected using an electrochemical detector with a glassy carbon electrode and an applied potential of 0.25 V. This same method can also be used to simultaneously separate and detect $Pb^{2+}$ and $Fe^{3+}$. The AS-5, like the AS-4, contains the 15-μm surface sulfonated cation exchange substrate. This substrate is coated with a monolayer of fully aminated latex beads of approximately 0.1 μm micrometer in diameter. The surface of the pellicular cation-exchange substrate is not fully coated with the anion exchange latex beads, however. As a result, there are some cation-exchange sites on the AS-5 to which cations such as Pb( + 2) and Fe( + 3) can bind. Thus, gold, platinum, palladium, lead, and iron can all be determined with the AS-5 column and UV detection.

The AS-5 column can also be used for simultaneous anion and cation determinations. The cations are first converted to anionic metal-EDTA complexes, and are separated from the other anions in the sample. The AS-5 column is preferred for this separation because the excess, unbound EDTA is a large hydrophobic anion. As a result, it is strongly retained on most anion separator columns. EDTA, though, is like most anions in that it is so strongly retained on the AS-5. Thus, the excess EDTA can be washed off the AS-5, permitting it to then be used for other separations. A simultaneous determination of acetate, nitrate, lead, nickel, sulfate, and copper is illustrated in Figure 10. The metals have been determined as their EDTA complexes. The determination of metal-EDTA complexes will be discussed in more detail in Chapter 4 (Section I).

Although the AS-5 has numerous advantages when separating large, hydrophobic anions, it has not appeared in the literature nearly so often as have other anion separator columns. This should change in the near future with the advent of gradient elution. Gradient elution is especially useful in bringing together peaks that are widely separated, by using an isocratic eluent. The AS-5 has special utility for gradient, because most all anions are less strongly retained on the AS-5. Thus, widely separated peaks can be brought closer together by (1) using the AS-5 column, on which anions elute faster, and (2) by using a gradient. The AS-5 achieves its special utility for anion separations because anions are less strongly retained on it. In general, it is easier to separate weakly retained anions because an eluent can always be made weaker, but strongly retained anions cause special problems. There are always limits to how strong an eluent can be made before it can no longer be suppressed. Thus, the fewer anions that are strongly retained, the better. In fact, fewer anions are strongly retained on the AS-5 than on the other anion separator columns. The AS-5 column may not be so well known in the previously published literature, but it soon will be.

The AS-4 (and AS-4A) provide rapid anion separations in a manner quite familiar to experienced ion chromatographers. Thus, it has been used in numerous published applications. The AS-5 provides the capability for separating large, hydrophobic anions and is now beginning to show special utility in gradient separations. There is still an important limitation that applies to the AS-4, 4A, and 5. These separators all have limited anion-exchange capacity. There is a limit to the concentration of eluent that can be used. High eluent

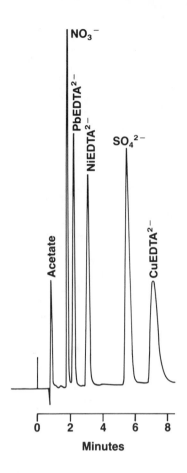

FIGURE 10. Separation of anions and metal EDTA complexes on an AS-5 column. A metal-free (Chelex) precolumn is needed to remove trace metals from the eluant.

concentrations that may be needed for NaOH gradient elutions can overload an AS-4, 4A, or 5. In another application, carbohydrates require a high concentration of hydroxide to exist as anions. It was for this application that the next anion separator, the AS-6, was developed.

The AS-6 has a 10-μm polystyrene/divinylbenzene core that is 5% cross-linked. The latex beads are larger than any of the previous (AS-1 through 5) anion separators. The R group on the $-NR_3$ is hydrophobic. The AS-6 has a higher ion exchange capacity than the other separators, permitting the use of more-concentrated eluents. This is needed for carbohydrates because they are separated as anions. To exist as anions, they must be in highly alkaline solution, i.e., 0.15 $N$ KOH. Only the AS-6 column has the needed anion-exchange capacity to handle this eluent. If the AS-1, 2, 3, 4, 4A, or 5 separators were used, they would be overloaded by the 0.15 $M$ KOH needed to convert the carbohydrates to anions. Also, the high anion-exchange capacity of the AS-6 became handy when the micromembrane suppressor was developed. Monoprotic organic acids can be separated using a 20-m$M$ sodium tetraborate eluent as shown in Figure 11. Divalent organics can also be separated, but they require the stronger 20-m$M$ sodium carbonate plus the 2-m$M$ NaOH eluent, as shown in Figure 12.

The high capacity of the AS-6 is also useful when analyzing the anionic content of strong electrolytes such as NaOH. Schibler et al.[23] used the AS-6 and the micromembrane suppressor to determine trace anionic contaminants in NaOH. Previous attempts to analyze sodium hydroxide used other techniques to reduce the ionic strength of the NaOH sample. This can

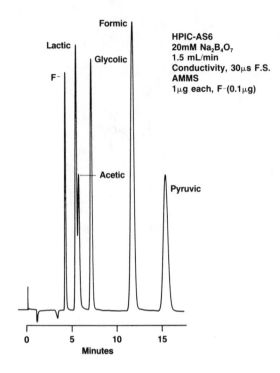

FIGURE 11.    Monoprotic organic acids by anion exchange. Column: AS-6; detection: conductivity with an AMMS.

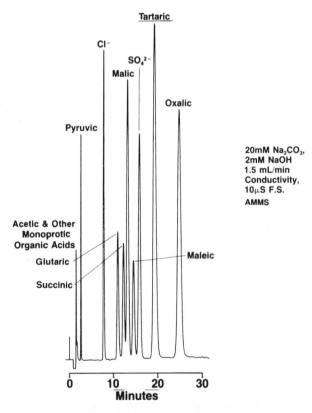

FIGURE 12.    Diprotic organic acids by anion exchange. Column: AS-6, detection: conductivity with an AMMS.

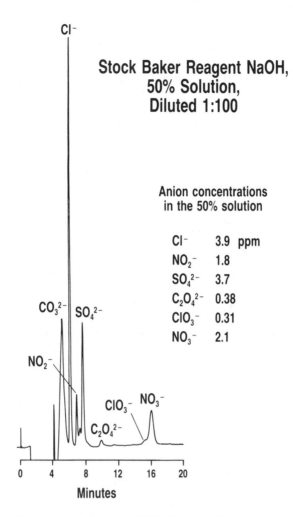

FIGURE 13.   Determination of anions in NaOH. Low detection limits are possible because the AMMS is used. The AMMS can suppress larger quantities of NaOH than can other suppressors, allowing the analyst to inject more concentrated NaOH samples.

be done by dilution[18] or by sample pretreatment with a cation exchange resin. The dilution procedure is fast and easy, but as the NaOH is diluted, so also are the anionic contaminants. As a result, the anions approach their detection limits. The sample pretreatment with cation exchange system does not alter the detection limits, but it is somewhat time-consuming, and is not amenable to automation. The procedure reported by Schibler using the AS-6, however, is rapid, amenable to automation, and has ppb detection-limits for the trace anions. A sample chromatogram illustrating the determination of trace anions in NaOH is illustrated in Figure 13. Chromatography was performed on a Model 2010i ion chromatograph which was plumbed in such a way as to optimize sensitivity. An auxiliary pump was used to load the sample from a 4-m$\ell$ sample loop onto an AG-6 concentrator column using deionized water. A second AG-6 column in the hydroxide form was used to trap any anions in the purge water. A metal-free column (MFC-1) was used as a metal trap column to remove any trace metallic contaminants which could interfere with the phosphate and oxalate determinations. Phosphate and oxalate are good chelating agents for metals. Thus, it was important to remove metals, or the phosphate and oxalate would be converted to metal complexes and would not be detected in the NaOH. It is also advantageous when determining trace levels of these and other potential chelating agents to be using the ion chromatograph which is made of nonmetallic, noncorrodable materials.

## III. ION CHROMATOGRAPHY EXCLUSION (ICE)

ICE is a completely different mechanism for separating anions. The ICE separators are not anion-exchange columns. Instead, they are porous, sulfonated polystyrene/divinylbenzene, and are technically cation-exchange resins. The typical eluent for ICE is 1 m$M$ HCl. At pH = 3, strong acids such as HCl, HNO$_3$, and H$_2$SO$_4$ are fully dissociated. The anions Cl$^-$, NO$_3^-$, and SO$_4^{2-}$ are fully dissociated and are excluded from the negatively charged pores on the ICE column and elute in the void volume. The fixed negative charge on the ICE column repels other negative charges. This phenomenon, called Donnan exclusion, is the basis for ICE. Weak acids, such as formic, acetic, and propionic, are only slightly dissociated and are only slightly repelled by the ICE column. These acids are retained on the ICE column because they are almost totally uncharged and can fit into the pores of the column. Thus, a strong acid which is fully ionized is excluded from an ICE column as shown in Figure 14. Depending on the pKa of the acid, it is either retained (if it has a high pKa) or it is excluded and elutes in the void volume (if it has a low pKa).

When comparing a homologous series of carboxylic acids, their retention-times increase as their pKa increases and their water solubility decreases. The pKa determines the degree of ionization of the acids. The higher the pKa, the less ionized the acid and the less it is excluded from the pores of the column. The water solubility is a measure of the polarity of the acid. Less-soluble acids, propionic acid vs. formic acid, are less polar and have stronger nonpolar interactions with the nonionic polystyrene/divinylbenzene. In fact, benzoic acids have a very strong interaction with the polystyrene/divinylbenzene and as such are strongly retained on the ICE column.

In general, dicarboxylic acids are eluted off an ICE column before their corresponding monocarboxylic acids, i.e., oxalic acid before acetic acid and malonic acid before propionic acid. The retention times of di- and tricarboxylic acids increase with increasing pKa. Double bonds tend to increase retention, i.e., acrylic acid elutes after propionic acid. From previously published data,[2,21] the elution order of several aliphatic carboxylic acids is as follows:

 1. Maleic
 2. Oxalic
 3. Citric acid
 4. Pyruvic acid
 5. Tartaric acid
 6. Malonic acid
 7. α-Ketobutyric acid
 8. α-Ketovaleric acid
 9. Succinic acid
10. Glyceric acid
11. Lactic acid
12. Formic acid
13. Adipic acid
14. Fumaric acid
15. Acetic acid
16. Propionic acid
17. Acrylic acid
18. Isobutyric acid
19. Butyric acid
20. Mandelic acid
21. 2,2-Dimethyl propionic acid
22. α-Hydroxy butyric acid

The first ICE-separator column, developed for the Model 10, 12, 14, and 16 ion chromatographs, was simply called the "normal" ICE separator and was capable of separations such as the one shown in Figure 15. The suppressor used when this chromatogram was obtained was the packed-bed "ICE suppressor". It was a cation-exchange resin in the silver form. The H$^+$ in the HCl eluent was exchanges for Ag$^+$ on the suppressor. The Ag$^+$ then combined with the Cl$^-$ in the eluent to form an AgCl precipitate which remained on the suppressor. Thus, both the highly conductive H$^+$ and Cl$^-$ were removed from the HCl eluent. After 1 to 2 days of use, the suppressor column had to be trimmed to remove the portion that had been saturated with AgCl.

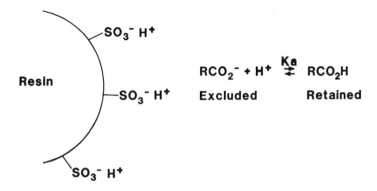

FIGURE 14. Ion exclusion separations. Strong acids (low pKa) are fully ionized, excluded, and elute in the void volume. Weak acids (high pKa) are partly ionized and are retained on the ICE column.

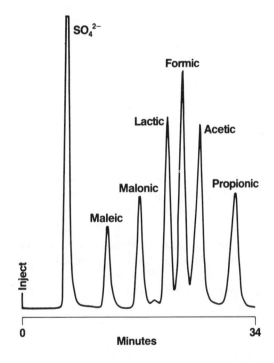

FIGURE 15. Standard ICE anion determination. Separation of organic acid anions on the first ICE column.

The next ICE column developed was a shorter column designed for faster analyses of low-pKa acids. One of the main applications for this column was the rapid separation of acids that inherently have very different affinities for the ICE column. Thus, if one needed to separate tartaric and acetic acids only, it would be unnecessary to use the ''normal'' ICE separator which required about 20 min per analysis. By using the low-pKa ICE column with its smaller void volume, the same separation could be done in about 10 min. This can be very useful in automated analysis of routine samples. By cutting the analysis time in half, twice the analyses can be performed in a completely unattended mode.

The first two ICE separators created limited back-pressure and could be used on the Models 10, 12, 14, and 16. With the introduction of the Series 2000 and high-performance ion chromatography (HPIC), the term ICE was changed to HPICE (for high-performance

ICE). A new, more efficient separator column was developed and is now called the HPICE AS-1. It has 7.5-$\mu$m polystyrene/divinylbenzene core particles 9% of which cross-linked. The ICE column for the Models 10, 12, 14, and 16 is now called the HPICE AS-2. A special low cross-linked resin for better separation of organic acids with low pKa (i.e., pKa 2 to 4) is called the HPICE AS-3. It has 15-$\mu$m polystyrene/divinylbenzene core particles with 4% cross linking. As with the previously discussed "low-pKa separator", the HPICE AS-3 is especially useful for rapid separation of organic acids that have very different affinities for ICE separators. The HPICE AS-3 can also be useful when combined in series with the HPICE AS-1. The two columns combined have more theoretical plates, which enables better separations.

Another advantage in the area of ion chromatography exclusion was the development of the second generation anion fiber suppressors, called the AFS-II. Before the introduction of the AFS-II, the packed-bed Ag-form suppressor was used with ICE separators. The packed-bed suppressors require periodic cutting to remove the portion that has become saturated with AgCl. The AFS-II, on the other hand, is capable of suppressing the conductivity of the HCl eluent without the need for any manual adjustments such as cutting or trimming. The fiber in the AFS-II is much more resistant to detergents such as octane sulfonic acid than the fiber in the AFS-I. As a result, octane sulfonic acid can be used as the regenerant.

The next advance in HPICE was the introduction of a fourth separator column especially designed for amino acid analysis, the HPICE AS-4. It has a 6-$\mu$m diameter polystyrene/divinylbenzene core particle with 9% cross linking. It is used to separate amino acids based on their differences in charge (pKa), polarity, and size. Instead of using chemically suppressed conductivity detection, they are detected by post-column reaction with either ninhydrin (for proline) or *o*-phthaldehyde (for primary amines). This is discussed in more detail in Chapter 3.

The next advance in HPICE was the development of a special mixed-mode HPICE separator called the HPICE AS-5. It is called a mixed-mode separator because ion exclusion is not the only mechanism, or mode, of retention. Unlike the other HPICE columns, the HPICE AS-5 retains carboxylic acids by hydrophobic interactions and hydrogen bonding. The HPICE AS-5 is also unique in that it is not simply a sulfonated polystyrene/divinylbenzene. Instead, it contains a mixture of strong and weak cation exchange sites on a polymeric, pH-resistant, substrate. For many years, ion exchange chromatographers have used the term "strong cation exchange site" to signify a sulfonate ($-SO_3H$) group and weak cation exchange site to signify a carboxyl ($-COOH$) group.

The HPICE AS-5 is similar to the other HPICE columns in that it is porous, and the pores do have some sulfonate groups attached. Thus, as with the other HPICE columns, strong, fully ionized acids (pKa < 2) are completely excluded and co-elute in the void volume. Similarly, in a homologous series of carboxylic acids (e.g., formic, acetic, and propionic acids), they elute in order of increasing pKa. Unlike the other HPICE columns, however, Donnan exclusion is only one of the retention mechanisms. This means that aliphatic carboxylic acids have a different elution order on the HPICE AS-5.

Because there are weak ion exchange sites (i.e., $-COOH$), the $-OH$ of the analyte acid can form a hydrogen bond with the C=O of the weak ion exchange site, and the C=O of the analyte acid can form a second hydrogen bond with the $-OH$ of the weak ion exchange site. Perhaps more important, the $-OH$ on hydroxy acids can also form hydrogen bonds with the weak ion-exchange sites of the HPICE AS-5. This was not possible on the HPICE AS-1. The hydroxy acids glucuronic, isocitric, citric, and galacturonic acids all co-eluted in one group. Ascorbic, malic, and quinic acid co-eluted in a second group. As shown in Figure 16, these hydroxy acids can be separated on the HPICE AS-5 column. The eluent was perfluorobutyric acid flowing at 0.3 m$\ell$/min. Enough perfluorobutyric acid was added to deionized water to give a pH of 2.8. The suppressor column used was the ICE anion

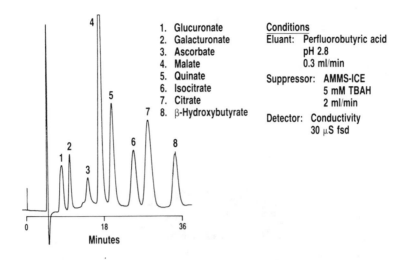

| | |
|---|---|
| 1. Glucuronate | |
| 2. Galacturonate | |
| 3. Ascorbate | |
| 4. Malate | |
| 5. Quinate | |
| 6. Isocitrate | |
| 7. Citrate | |
| 8. β-Hydroxybutyrate | |

**Conditions**

Eluant: Perfluorobutyric acid
pH 2.8
0.3 ml/min

Suppressor: AMMS-ICE
5 mM TBAH
2 ml/min

Detector: Conductivity
30 μS fsd

FIGURE 16.    Hydroxy organic acids on the newer, more efficient HPICE-AS-5.

micromembrane suppressor. It was continuously regenerated with tetrabutylammonium hydroxide (TBAOH) at 2 mℓ/min.

There are several reasons for choosing an eluent such as perfluorobutyric acid instead of the 1 m$M$ HCl used so often with the other HPICE columns. To begin with, the continuously regenerated anion micromembrane suppressor was used instead of the disposable packed-bed suppressor. The packed-bed suppressor is in the silver form, so it would convert the HCl eluent to AgCl and H$_2$O. Unfortunately, the packed-bed suppressor had to be manually trimmed periodically, limiting the level of automation attainable. With a micromembrane suppressor that is continuously regenerated with TBAOH, the HCl eluent would be converted to tetrabutylammonium chloride, which still has significant conductivity. On the other hand, eluents such as perfluorobutyric acid and octanesulfonic acid are converted to the much less conductive tetrabutylammonium perfluorobutyrate or octanesulfonate. Both of these salts of tetrabutylammonium hydroxide are only slightly dissociated and therefore are only slightly conductive. The suppression reaction that occurs in the ICE micromembrane suppressor is illustrated in Figure 17. The eluent (generalized as HA) enters at one end of the suppressor, and the regenerant (TBAOH) enters at the other end and flows countercurrent to the eluent. At the end in which eluent enters, the suppressor is fully expended and is, therefore, in the H$^+$ form. At the opposite end where the regenerant enters, the suppressor is fully regenerated and is in the TBA form. Between these two extremes in a region where there is a mixture of expended and regenerated sites, i.e., H$^+$/TBA$^+$. The membrane is semipermeable, so that the anion from the eluent (perfluorobutyrate, when perfluorobutyric acid is used) can form an ion pair (TBA$^+$/A$^-$) which goes to waste. Thus, suppression of the eluent is achieved.

Using the perfluorobutyric acid eluent, the HPICE AS-5 can separate up to 12 carboxylic acids as shown in Figure 18. Although chloride is shown as the first peak, other strong, inorganic acids (nitric, sulfuric) would co-elute in the void volume. It is significant that tartarate can be separated from citrate and that lactate, malate, acetate, and succinate can be separated. This was not possible with the other HPICE columns. Chromatography of hydroxy acids has always been difficult. Many carboxylic acids without hydroxy groups can be derivatized with reactive silicone compounds (silanizing agents) to form volatile compounds that are suitable for gas chromatography. This is a commonly used method for determining free fatty acids in biological samples.[24] Hydroxy acids do not readily form volatile derivatives, so the HPICE AS-5 represents a significant development.

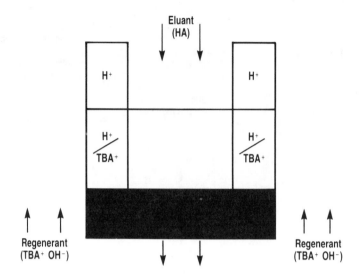

FIGURE 17.   Suppression reaction in an anion micromembrane suppressor for ICE. The regenerant in this example is tetrabutyl ammonium hydroxide (TBAOH).

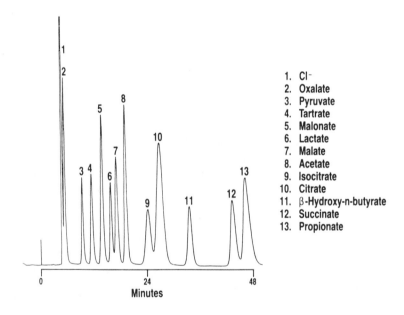

1. Cl⁻
2. Oxalate
3. Pyruvate
4. Tartrate
5. Malonate
6. Lactate
7. Malate
8. Acetate
9. Isocitrate
10. Citrate
11. β-Hydroxy-n-butyrate
12. Succinate
13. Propionate

FIGURE 18.   Organic acids on HPICE-AS-5. Note improved efficiency over the original ICE column.

The capacity factors (k′) for 28 different anions of weak acids are shown in Table 5. It should be noted that just because an anion is inorganic (such as fluoride and phosphate), it can be retained on an ICE column, provided that it has a high enough pKa (pKa for HF is 2.7 and for $HPO_4^{2-}$ it is 3.2). It should also be noted that some acids which co-elute can be separated by the addition of 1% isopropanol solvents to the perfluorobutyric acid eluent. This is because of a third mechanism, or mode, of retention, i.e., hydrophobic interactions. The hydrophobic alkyl portions of fumarate, propionate, and acetate interact with the hydrophobic polymeric core of the HPICE AS-5, and are more strongly retained. This mech-

**Table 5**
**K′ VALUES AT PH 2.8**

| Analyte | K′ | Analyte | K′ |
|---------|-----|---------|-----|
| F$^-$ | 0.2 | Malate | 2.6 |
| $H_2PO_4^{2-}$ | 0.2 | Acetate | 2.9[a] |
| Oxalate | 0.2 | Quinate | 3.3 |
| Glucuronate | 0.7 | cis-Aconitate | 3.5 |
| Gluconate | 1.1 | α-Ketoglutarate | 3.8 |
| Galacturonate | 1.1 | Hydroxyisobutyrate | 3.8 |
| Pyruvate | 1.0 | Isocitrate | 4.1 |
| Tartrate | 1.4 | Citrate | 4.8 |
| Formate | 1.4[b] | β-Hydroxybutyrate | 6.3 |
| Glycolate | 1.4 | Hydroxyglutarate | 6.4 |
| Malonate, | 1.7 | α-Hydroxyvalerate | 6.9 |
| $HCO_3^-$ | | | |
| Ascorbate | 1.9 | Succinate | 8.3 |
| Lactate | 2.2 | Propionate | 9.2[c] |
| Maleate | 2.3 | Fumarate | 10.5[d] |

[a]   2.3 (1% solvent).
[b]   1.3 (1% solvent).
[c]   4.8 (1% solvent).
[d]   7.4 (1% solvent).


**Table 6**
**K′ DEPENDENCE ON pH HPICE-AS5**

| Acid | pKa | K′ vs. pH | | | |
|------|-----|--------|------|------|------|
| | | pH 2.2 | 2.6 | 2.8 | 3.8 |
| Tartaric | 2.96, 4.24 | 2.04 | 1.50 | 1.24 | 1.15 |
| Formic | 3.76 | 2.04 | 1.50 | 1.38 | 1.38 |
| Lactic | 3.86 | 2.42 | 2.33 | 2.25 | 2.17 |
| Malic | 3.40, 5.05 | 3.17 | 2.83 | 2.63 | 2.38 |
| Acetic | 4.76 | 3.17 | 3.08 | 3.00 | 2.96 |
| Citric | 3.13, 4.76, 6.40 | 7.33 | 5.83 | 5.13 | 4.88 |

anism of retention can be partly negated by adding the relatively hydrophobic isopropanol. Isopropanol is more hydrophobic than the acids, so it preferentially binds to the available hydrophobic sites on the HPICE AS-5. This blocks the hydrophobic interactions between the carboxylic acids and the HPICE AS-5, thus decreasing retention (and k′). It is important to note that all this data is for eluent pH of 2.8. Resolution can also be controlled by adjusting the eluent pH, as shown in Table 6. As the pH of the eluent increases, the carboxylic acids become more ionized, and are excluded more (by Donnan exclusion). This means that the carboxylic acids are not so strongly retained as the pH increases. The most notable separation that can be achieved by changing eluent pH is between tartarate and formate. Similarly, lactate, malate, acetate, and citrate resolution, is also affected by pH.

There are several applications using the AS-5. The wine industry has been interested for some time in monitoring lactate/malate ratios during fermentation. As mentioned previously, these hydroxy acids are not easily derivatized for gas chromatography. Other attempts at carboxylic acid determinations using ion exchange[25] or reversed-phase HPLC[26] suffered from the lack of a chemically suppressed conductivity detection. Using the HPICE AS-5, however, wine need only be diluted, filtered, and injected to effectively determine lactate and malate simultaneously in one run.

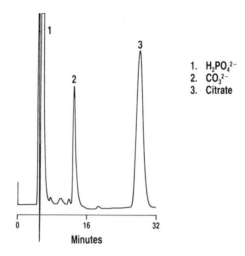

1. $H_2PO_4^{2-}$
2. $CO_3^{2-}$
3. Citrate

FIGURE 19. Determination of weak acids in Diet Coke® on HPICE-AS-5. Sample preparation, dilute, filter and inject. The peak at 14 min is due to carbonate.

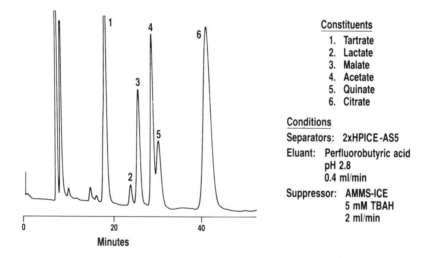

**Constituents**

1. Tartrate
2. Lactate
3. Malate
4. Acetate
5. Quinate
6. Citrate

**Conditions**

Separators: 2xHPICE-AS5

Eluant: Perfluorobutyric acid
pH 2.8
0.4 ml/min

Suppressor: AMMS-ICE
5 mM TBAH
2 ml/min

FIGURE 20. Determination of weak acids in coffee on HPICE-AS-5.

High-citrate foods make up another major application area for this separator. Pyruvate, tartarate, ascorbate, malate, acetate, isocitrate, citrate, and succinate were determined simultaneously in one injection of a 1:5 dilution of filtered tomato juice. Similarly, phosphate, carbonate, and citrate can be determined in diet Coke as shown in Figure 19.

Coffee manufacturers try to monitor malate/quinate levels, which can affect its taste. With other HPICE columns, these co-eluted. Using two HPICE AS-5 separators in series, the separation shown in Figure 20 can be achieved.

Although several examples have been given that illustrate the separation of many carboxylic acids in one run, not all samples contain so many components. It should be remembered that carboxylic acids can be made to elute faster by either raising the eluent pH or by adding 1% propanol. For example, if the sample contains only chloride, formate, acetate, and proprionate, it would take 48 min per analysis if the chromatographic conditions from Figure 18 were used. Alternatively, these four anions can be separated in 16 min if 1% isopropanol is contained in the eluent.

A combination of HPICE and ion exchange chromatography can be a very powerful tool for qualitative analysis. For example, if a microbiologist is studying the metabolism of bacteria, it might be obvious from gas chromatographic analysis that some previously un-identified metabolites are present. It is reasonable to expect that a prior knowledge of the bacteria indicates that carboxylic acid metabolites are expected. Because the retention of carboxylic acids is so different by ion exchange (i.e., AS-6, Figures 11 and 12), HPICE AS-1 (Figure 15), and HPICE AS-5 (Figure 18 and Table 5), the chromatography of the filtered growth medium can be compared to that of standards on three different columns. As in any chromatographic method, once the suspect carboxylic acids are identified, the filtered bacterial growth medium should be spiked. If the suspect-peaks from the ion chro-matographs co-elute with the added carboxylic acids on all three chromatograms, their identity can be reasonably assured. It would probably be best to put an MPIC guard column in front of the AS-6, HPICE AS-1, and AS-5 columns to prevent contaminating the valuable separators. The case of identifying new metabolites is just one example. Because there are at least three different separation modes available for aliphatic carboxylic acids, qualitative analysis is more possible for a variety of applications.

As shown by Rocklin et al.,[27] carboxylic acids can be separated by anion exchange on an AS-6 and a micromembrane suppressor. Diprotic acids in wine were separated on an AS-6 using 20 mM sodium carbonate plus 2 mM NaOH. Strongly retained diprotic and triprotic acids were separated on an AS-4 using this same eluent. As mentioned previously, the AS-6 has a higher anion exchange capacity, causing the triprotic acids (and others) to be more strongly retained. Microgram quantities of carboxylic acids were separated on an HPICE AS-1 and detected by chemically suppressed conductivity (an anion fiber suppressor-2 was used). An increase in acetic acid concentration in a fermentation broth was detected by HPICE. Citric, pyruvic, and lactic acids were determined in milk. Nitric, hydrofluoric, and acetic acids were determined in a mixed acid etchant by ICE.

## IV. CATIONS

The first cation separator, the CS-1, has a polystyrene/divinylbenzene core particle with a 20-μm diameter and is 2% cross-linked. This column was designed initially for the separation of monovalent cations lithium, sodium, ammonium, potassium, rubidium, and cesium. With the development of the cation-fiber suppressor, these six cations could be separated in 18 min using a 5-mM HCl eluent. The divalent alkaline earth metals require the much stronger eluent, 2.5 mM m-phenylenediamine · 2HCl plus 2.5 mM nitric acid. Using this eluent, magnesium, calcium, barium, and strontium are well separated, but the monovalent cations all co-elute in the void volume. This eluent is so strong that it cannot be washed of the CS-1 using the 5 mM HCl eluent required for monovalents. As a result, two separate CS-1 columns were required for mono- and divalent cation analysis. In spite of this, the CS-1 achieved great popularity for metals analysis. It offers a suitable alternative to atomic spectroscopy methods. If a laboratory is required to analyze anions anyway, it is significantly cheaper to obtain one more column (the CS-1) to analyze cations than to obtain a whole new instrument (such as an atomic absorption spectrophotometer). In addition, the CS-1 was the only separator column available for metals analysis for a number of years. As a result, it is the most well-known and most thoroughly described cation separator in the literature.

One of the reasons for its popularity is that the CS-1 can be used for other analyses. Although it was not originally designed with aliphatic amines in mind, several laboratories have described their separation on the CS-1. Reutter[28] reported the determination of mon-omethylamine in bombing investigations. They used an eluent consisting of 5 mM HCl with 40% methanol flowing at 2 mℓ/min. The reported retention times were 12 min for sodium, 14 min for ammonium, 18 min for monomethylamine, and 21 min for potassium. The

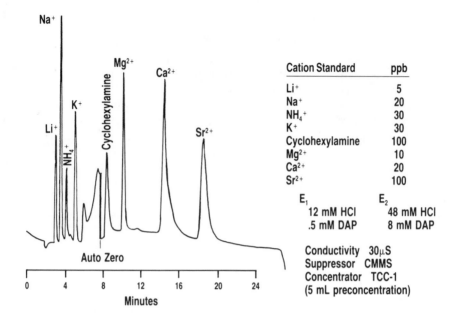

The following data accompanies the figure:

| Cation Standard | ppb |
| --- | --- |
| Li+ | 5 |
| Na+ | 20 |
| NH4+ | 30 |
| K+ | 30 |
| Cyclohexylamine | 100 |
| Mg2+ | 10 |
| Ca2+ | 20 |
| Sr2+ | 100 |

| E1 | E2 |
| --- | --- |
| 12 mM HCl | 48 mM HCl |
| .5 mM DAP | 8 mM DAP |

Conductivity 30μS
Suppressor CMMS
Concentrator TCC-1
(5 mL preconcentration)

FIGURE 21.    Simultaneous trace monovalent and divalent cations. Column: CS-3 (step gradient).

presence of methanol does not affect the retention of sodium and ammonium, but is needed to separate potassium from monomethylamine. Without methanol, they co-elute at about 16 min.

In reports by Buechele and Reutter,[29] ethylenediamine was eluted off the CS-1 with an eluent consisting of 4 m$M$ HCl plus 2.5 m$M$ ZnCl$_1$. All the alkali metals and the following monoamines co-eluted in the void volume: methylamine, trimethylamine, ethylamine, diethylamine, triethylamine, ethanolamine, diethanolamine, and triethanolamine. Bouyoucos[25] found that dimethylamine and trimethylamine could be eluted after ammonium using a 10 m$M$ HCl eluent. In a different study, Gilbert determined morpholine and cyclohexylamines as cations, separated on a CS-1.[31] Volatile amines such as morpholine and cyclohexylamine are used as corrosion inhibitors in the stem-water cycle in fossil fuel and nuclear power plants. Another cation, hydrazine, is used as an oxygen scavenger and acts to passivate metal surfaces. With the standard 5 m$M$ HCl eluent, morpholine eluted in about 26 min (ammonium in 12 min). Cyclohexylamine requires the stronger eluent, 3 m$M$ HCl plus 2.5 m$M$ lysine, and elutes in about 10 min (sodium, ammonium, potassium, and morpholine all co-elute at about 5 min). To determine hydrazine, this same lysine/HCl is used, but hydrazine is detected using the electrochemical detector. In a similar study, Saitoh et al.[32] determined choline and tetramethyl ammonium, a cause of food-poisoning in shell fish, but they used the second cation separator, the CS-2, and a 10-m$M$ HCl eluent.

This CS-2 column was actually developed for separating transition metals. The monovalent alkali metal cations and aliphatic amines are more strongly retained on the CS-2 than on the CS-1. Transition metals co-elute between sodium and ammonium on the CS-1 using an HCl eluent. The transition metals are then precipitated as their hydroxides when they reach the suppressor column. Transition metal analysis is discussed in more detail in Chapter 3 (Section I).

A significant advance in the determination of alkali and alkaline earth-metal cations occurred with the development of the CS-3 column. It has a higher cation exchange capacity than the CS-1 and is more efficient (i.e., it has a higher height equivalent per theoretical plate, HETP). Mono- and divalent cations can be determined simultaneously in one run. This was done originally using a single step gradient as shown in Figure 21. The eluent starts out as 10 m*M* HCl plus 0.09 m*M* zinc chloride plus 0.19 m*M* diaminopropionic acid (DAP) and is switched to 40 m*M* HCl plus 1.5 m*M* zinc chloride plus 3 m*M* DAP after 3 min. With the cation micromembrane suppressor, this strong eluent is suppressed and there is little baseline drift when the eluent changes. These stronger eluents are required because cations are more strongly retained on the CS-3. Because it has not been available very long, the CS-3 has not yet achieved the notoriety that the CS-1 has. In spite of this, the CS-3 is most probably the better column to use for all applications in which alkali metals, ammonium, and/or alkaline earth metals are to be determined.

# REFERENCES

1. **Small, H., Stevens, T. S., and Bauman, W. C.,** Novel ion exchange method using conductimetric detection, *Anal. Chem.,* 47, 1801, 1975.
2. **Weiss, J.,** *Handbuch der Ionenchromatographie* (Handbook of Ion Chromatography)., Friederich Ehrenklau Druckerei GmbH, Hessen, W. Germany, 1985.
3. **Schiff, L. J., Pleva, S. G., and Sarver, E. W.,** Analysis of phosphonic acids by ion chromatography, in *Ion Chromatographic Analysis of Environmental Pollutants,* Vol. 2, Ann Arbor Science, Ann Arbor, Mich., 1977.
4. **Mirna, A., Wagner, H., Klotzer, E., and Fausel, E.,** Zur Anwendung der Ionenchromatographie bei der Untersuchung von Fleisch und Fleischwaren (Applications of ion chromatography in the study of meat and meat products), *Lebensmittelchem. Gerichtl. Chem.,* 38, 18, 1984.
5. **Lash, R. P. and Hill, C. J.,** Ion chromatographic determination of dibutylphosphoric acid in nuclear reprocessing streams, in *Ion Chromatographic Analysis of Environmental Pollutants,* Vol. 2, Ann Arbor Science, Ann Arbor, Mich., 1977.
6. **Deister, H. and Runge, E. A.,** Ehrfahrungen mit der Ionenchromatographie in einem Laboratorium der Eisen und Stahlindustrie (Experience with ion chromatography in a laboratory of the iron and steel industry), *Arch. Eisenhuttenwes.,* 54, 405, 1983.
7. **Lindgren, M., Cedergren, A., and Lindberg, J.,** Conditions for sulfite stabilization and determination by ion chromatography, *Anal. Chim. Acta,* 141, 279, 1982.
8. **Bodek, I. and Smith, R. H.,** Determination of ammonium sulfamate in air using ion chromatography, *Am. Ind. Hyg. Assoc. J.,* 41, 603, 1980.
9. **Ficklin, W. H.,** The separation of tungstate and molybdate by ion chromatography and its application to natural waters, *Anal. Lett.,* 15, A10, 865, 1982.
10. **Chakraboti, D., Hillman, D. C. J., Irgolic, K. J., and Zingaro, R. A.,** Hitachi Zeeman graphite furnace atomic absorption spectrometer as a selenium-specific detector for ion chromatography, *J. Chromatogr. Sci.,* 18, 442, 1980.
11. **Pohl, C. and Edward, L.,** Ion chromatography — the state of the art, *J. Chromatogr. Sci.,* 18, 442, 1980.
12. Dionex Corporation, Trace sulfate and phosphate in brine, Application Note 3, 1978.
13. **Tan, L. K. and Dutrizak, J. E.,** Determination of arsenic (III) and arsenic (V) in feric chloride-hydrochloric acid leaching media by ion chromatography, *Anal. Chem.,* 57, 1027, 1985.
14. **Hoover, T. B. and Yager, G. D.,** Comparison of collection procedures for the reinjection ion chromatography of water, *J. Chromatogr. Sci.,* 22, 435, 1984.
15. **Hoover, T. B. and Yager, G. D.,** Comparison of trace anions in water by multidimensional ion chromatography, *Anal. Chem.,* 56, 221, 1984.
16. **Phillipy, B. Q. and Johnston, M. R.,** Determination of phytic acid by ion chromatography with postcolumn derivitization, *J. Food Sci.,* 50, 541, 1985.
17. **Cox, J. A. and Tanaka, N.,** Determination of anionic impurities in selected concentrated electrolytes by ion chromatography, *Anal. Chem.,* 57, 383, 1985.
18. **Smith, R. E. and Davis, W. R.,** Determination of chloride in sodium hydroxide and in sulfuric acid by ion chromatography, *Anal. Chem.,* 55, 1427, 1983.
19. **Müller, K. P.,** Ionenchromatographie in Niederschlagwasser (Ion chromatography in precipitation), *Fresnius Z. Anal. Chem.,* 317, 345, 1984.
20. **Smith, R. E. and Davis, W. R.,** Determination of chloride and sulfate in chromic acid by ion chromatography, *Plating Surf. Finish.,* 71, 60, 1984.
21. **Smith, R. E. and Smith, C. H.,** Automated ion chromatography — applications to the analysis of plating solutions, *L.C. Mag.,* 3, 578, 1985.

22. **Rocklin, R. D.,** Determination of gold, palladium, and platinum at the parts per billion level by ion chromatography, *Anal. Chem.,* 56, 1959, 1984.

23. **Schibler, J. A., Rocklin, R. D., and Rubin, R. B.,** Determination of trace anionic components in sodium hydroxide by ion chromatography, in preparation.

24. **Su, K. L. and Sun, G. Y.,** Acyl group composition of metabolically active lipids in brain: variances among subcellular fractions and during postnatal development, *J. Neurochem.,* 31, 1043, 1978.

25. **Bouyococos, S. A.,** Determination of organic anions by ion chromatography using a hollow fiber suppressor, *J. Chromatogr.,* 242, 170, 1982.

26. **Manning, D. L. and Maskarinec, M. P.,** Determination of low molecular weight carboxylic acids in water by HPLC with conductivity detection, *J. Liq. Chromatogr.,* 6, 705, 1983.

27. **Rocklin, R. D., Pohl, C. A., and Schibler, J. A.,** Gradient elution in ion chromatography, in preparation.

28. **Reutter, D. J., Buechele, R. C., and Rudolph, T. L.,** Ion chromatography in bombing investigations, *Anal. Chem.,* 55, 1468A, 1983.

29. **Buechele, R. C. and Reutter, D. J.,** Determination of ethylenediamine in aqueous solution by ion chromatography, *Anal. Chem.,* 54, 2113, 1982.

30. **Gilbert, R., Rioux, R., and Saheb, S. E.,** Ion chromatographic determination of morpholine and cyclohexylamine in aqueous solutions containing ammonia and hydrazine, *Anal. Chem.,* 56, 106, 1984.

31. **Saitoh, H., Oikawa, K., Takano, T., and Kamimura, K.,** Determination of tetramethylammonium ion in shellfish by ion chromatography, *J. Chromatogr.,* 281, 397, 1983.

Chapter 3

# NONCONVENTIONAL METHODS

In the first few years after the invention of ion chromatography, the only separation methods used were ion exchange, and the only detection method was chemically suppressed conductivity. This combination is so useful in solving a number of analytical problems that it is often forgotten that other separation and detection methods are available. Indeed, the original definition of ion chromatography did imply the use of suppressor columns. This definition is reinforced by the fact that most publications in the field of ion chromatography still describe the use of ion-exchange columns and conductivity detectors. The most well-known methods include the separation and detection of common anions and monovalent cations in aqueous samples. Certainly, ion chromatography deserves the reputation it has for rapid, easy determination of anions and cations based on only slight modifications of the original invention.

On the other hand, there are many other applications which either do not use ion exchange separators or chemically suppressed conductivity detection. These applications can still be considered to be in the realm of ion chromatography if the original definition is only slightly expanded. It is perhaps better to define ion chromatography as the chromatography of ions. In this way, the separation and detection methods are not specified. Columns and detectors can be mixed and matched to optimize the determination of the ion or ionic compounds of interest. Electroactive species can be separated as ions on ion-exchange separators and detected with an amperometric detector.

Large, hydrophobic ions, especially organic ions, can be separated as ion pairs on an uncharged column and still be detected by chemically suppressed conductivity. Still other ions, such as transition metals and lanthanide metals, can be separated, based on their differential affinities for chelating agents and be detected by post-column reaction with a colorimetric reagent. Probably the most important class of ions, amino acids, can also be separated and detected using an ion chromatograph. In fact, completely nonionic organic compounds in aqueous solutions can be separated and detected using unique ion chromatography hardware that is not available on any other analytical instrument.

Thus, the definition of ion chromatography is rapidly becoming muddled. It can even be used to determine nonionic compounds. For example, an ion chromatograph is the instrument of choice for analyzing highly corrosive plating and etching baths for nonionic organics. The nonmetallic components available in an ion chromatograph, together with chemically inert polystyrene/divinylbenzene separator columns are not available in other instruments. Whether this is truly an area of ion chromatography or reversed-phase liquid chromatography is a debate best left to the field of patent law. The important point is that these applications have not been described in any detail in the previous literature. Since these applications can all be performed with the same instrumental hardware that is marketed as an ion chromatograph, it is appropriate that the analyst be aware of the subject. In fact, papers are constantly being presented at ion chromatography conferences that describe the use of an ion chromatograph to determine organics that were simply not envisioned in the original invention. It is important to know, however, that an electroplater or an environmental scientist can use one instrument, an ion chromatograph, to thoroughly analyze their samples for a wide variety of components.[1]

The purpose of this chapter, then, is to expose the reader to alternative separation and detection methods that probably were not envisioned in the original invention. Some of these separation and detection methods are based on well-known principles used in more classical liquid chromatography. In every case, however, significant advances have been made that were deserving of new patents. Thus, a new membrane post-column reactor provides mixing of analytes with reagent, with a minimum increase in volume. Certainly, other post-column reactors have been used for years, before ion chromatography was even invented. They usually incorporated a mixing "T" in which the analytes mix with the post-column reagent. The resulting increase in volume caused band spreading and loss of sensitivity. The post-column membrane reactor, on the other hand, allows reagent and analyte to mix across a semipermeable membrane, resulting in a minimum volume increase.

In another example, significant improvements have been made in carbohydrate analysis. Certainly, carbohydrates have been separated by liquid chromatography on nonionic columns, but the lack of a significant chromophore prevents their sensitive detection by the UV detector that is standard on most liquid chromatographs. Thus, it was a significant advance when a pulsed amperometric detector became available to detect subparts per million levels of carbohydrates. It was also uniquely advantageous to separate the carbohydrates as anions. Usually, simple carbohydrates are not considered to be ionic. However, at very high pH, they become anions. Thus, it was significant when the AS-6 column with its higher ion-exchange capacity was developed to enable the use of NaOH eluents as concentrated as 0.16 $M$; carbohydrates exist as anions in this highly alkaline eluent. Thus, new patents are constantly being awarded in the name of ion chromatography because they describe novel methods for separating and detecting ionic substances.

## I. POST-COLUMN REACTION WITH UV-VISIBLE OR FLUORESCENCE DETECTION

One of the first limitations that became apparent with the original invention was that suppressor columns cannot be used in transition or lanthanide metal analysis. Transition metals cannot be detected by chemically suppressed conductivity because they form insoluble hydroxides on the suppressor. The cation suppressor converts all metals to their hydroxide form. Although this is useful in detecting lithium, sodium, ammonium, potassium, and alkaline earth metals, it means that it cannot be used for transition metals or lanthanide metals. To detect transition metals, then, it is necessary to use another technique. Fortunately, the transition metals react rapidly with a spectrophotometric reagent, 4-(2-pyridylazo) re-sorcinol (PAR) to form a highly colored complex. These complexes all absorb at approximately the same wavelength, in the region of 520 nm. These two facts make PAR an ideal reagent for the detection of transition metals. The metal-PAR complex is formed through the diazo nitrogens and the anionic phenolate on the PAR, which contains two acidic hydrogens due to its two phenolic groups. The complexes with metals are formed more rapidly when the PAR carries a negative charge. Thus, to detect the transition metals, the PAR is kept at about pH 9.7. At this pH, PAR exists primarily as a monovalent anion and is often abbreviated as HL. It forms a 1:1 complex with most metals at this pH.

Because of the metal-PAR complexes, all absorb in about the same region of the visible spectrum, the UV-visible detector can be set at one wavelength, 520 nm, and each metal-PAR complex can be detected. The PAR is mixed with the metals in a post-column membrane reactor. This post-column reactor represented a significant advance, because there is only a miniscule increase in volume when the analytes mix with the PAR reagent. The effluent off the CS-2 column enters at the top of the post-column reactor and flows inside the membrane, whereas the PAR enters at the bottom of the reactor and bathes the outside of the membrane. Thus, the PAR flows in the opposite direction to the CS-2 effluent. The PAR crosses the membrane and reacts with the metals as they are eluted off the CS-2. Because the metals and PAR mix across a membrane and not in a relatively large volume mixing "T", there is no detectable band spreading; this helps to optimize sensitivity. After mixing across the membrane, the colored metal-PAR complex then flows to the UV-visible detector. This represents a rapid and efficient method for detecting transition metals.

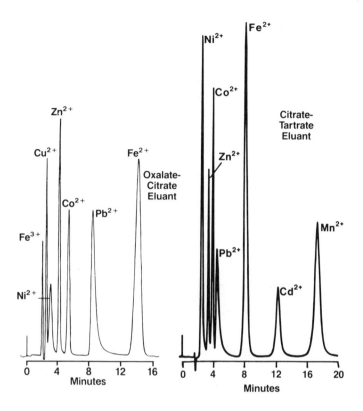

FIGURE 1. Separation of transition metals with HPIC-CS-2 column. Detection is by post-column reaction with PAR and visible detector.

For separation, the CS-1 column is not used because the transition metals are not very strongly retained. The CS-2 has a stronger affinity for the transition metals. In fact, the transition metals have very similar size and enthalpies of hydration, causing them to have very similar affinities for any cation-exchange resin. Thus, they cannot simply be eluted using an HCl eluent.

The transition metals do, however, have markedly different affinities for chelating agents. Thus, they are eluted off the CS-2 column with mild chelating agents. The mixture of chelating agents used, together with the pH of the eluent can be adjusted to alter the retention-times of the transition metals. For example, seven transition metals can be separated using 10 m$M$ oxalic acid plus 7.5 m$M$ citric acid adjusted to pH 4.3 with LiOH as seen in Figure 1. As with most chelating agents, oxalic acid and citric acid bind metals only when ionized. By raising the pH of the eluent, the transition metals elute faster. The other way to strengthen the eluent, of course, is to increase the eluent concentration or use stronger chelating agents such as tartaric acid. If the 10 m$M$ oxalic acid/7.5 m$M$ citric acid, pH 4.3 eluent from Figure 1 is used, $Cd^{2+}$ and $Mn^{2+}$ are strongly retained. To elute them, the stronger 40 m$M$ tartaric acid plus 12 m$M$ citric acid adjusted to pH 4.3 with LiOH is used. With this stronger eluent, however, the early eluting metals, $Fe^{3+}$, $Cu^{2+}$, and $Ni^{2+}$, would co-elute. Thus, the analyst must consider which transition metals need to be separated in order to decide which eluent to use. To verify the accuracy of the ion-chromatographic method for determining transition metals, two water samples were analyzed by atomic spectroscopy and ion chromatography. The results are summarized in Table 1. The detection limits for copper, zinc, and nickel are lower for ion chromatography than for atomic absorption (AA) and inductively coupled plasma (ICP) spectroscopy. The results obtained by ion chromatography and ICP were within 1 ppm (2% difference) for lead. The copper peak produced by sample 4668 was off-scale

## Table 1
## COMPARISON DATA

| | Limit of detection | | | Sample[a] | | |
|---|---|---|---|---|---|---|
| Pollutant | IC | AA | ICP | IC | AA | ICP |
| Lead (Pb) | 5 | 1 | 40 | 10 | 6 | ND |
| | | | | 53 | 63 | 54 |
| Copper (Cu) | 0.3 | 4 | 5 | 53 | 37 | 40 |
| | | | | O.S. | 6200 | 6000 |
| Cadmium (Cd) | 3 | 0.5 | 9 | 56 | 13 | ND |
| | | | | 570 | 530 | 500 |
| Zinc (Zn) | 3 | 15 | 4 | 500 | 600 | 560 |
| | | (flame) | | 686 | 790 | 830 |
| Nickel (Ni) | 3 | 75 | 5 | 25 | ND | 22 |
| | | (flame) | | 750 | 950 | 940 |

*Note:* Values are in ppm; IC, Ion Chromatography; AA, Atomic Absorption; ICP, Inductively Coupled Plasma; O.S., Off-Scale.

[a]  First values listed are for sample #4661; second values are for sample #4668.

on the ion chromatograph, but the sample could have been analyzed by simply diluting it. Certainly more extensive statistics must be collected before the methods can be properly compared for equivalency, but for many applications, these comparison data are quite encouraging.

Although PAR is certainly the most popular, there is no reason why other spectrophotometric reagents cannot be used for metal analysis. For example, Yan and Schwedt[2] used a reagent consisting of 0.01% chrome azurol S plus 0.01% cetyltrimethyl ammonium bromide (CTAB) and 0.05% Triton® X-100 for detecting aluminum and iron. They set the UV-visible detector at 610 nm and reported a detection limit of 2 ppb for aluminum. The eluent used was 10 m$M$ sulfosalicylic acid plus 2 m$M$ ethylenediamine at pH 5. The aluminum eluted in 3 min, and the ferrous ion in about 6 min. These same workers also reported the detection of bismuth, calcium, and magnesium, along with the transition metals from Figure 1, using post-column reaction with PAR. The other post-column reagent used was Arsenazo I, which binds calcium and magnesium to form a highly colored complex that can be detected at 570 nm.

It is also possible to vary the eluent to achieve different selectivity and enhanced analytical capabilities. The most important parameter that can be changed is the type of chelating agent in the eluent. It has been shown that pyridine-2,6-dicarboxylic acid (PDCA) is a very useful chelating agent for metals determinations.[3] In addition, a new cation-separator column, the CS-5, has been developed to enhance metals analysis. The CS-5 column has an efficiency of about 12,000 theoretical plates per meter. It is a novel form of cation separator, because it uses a special latex agglomerated onto a 13-μm substrate. All the agglomerated latex columns discussed so far are anion separators, where the latex was agglomerated onto a negatively charged substrate. The latex carried positive charges in the form of quatenary amines (latex-$NR_3$). The situation is just reversed in the CS-5. With this column, the latex beads have negative charges and it is the substrate that has a positive charge. Thus, the negatively charged latex beads attract cations.

This new CS-5, when used with PDCA, has different selectivity than the CS-2 as shown in Figure 2. With the CS-5 and PDCA, $Pb^{2+}$ elutes earlier, and $Mn^{2+}$ later, than on the CS-2. The $Pb^{2+}$ elutes before $Fe^{3+}$, $Cu^{2+}$, $Ni^{2+}$, $Zn^{2+}$, and $Co^{2+}$ with the CS-5, but $Pb^{2+}$

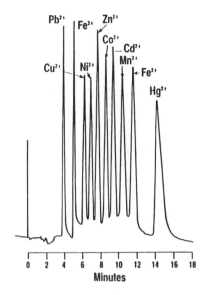

FIGURE 2.    Selectivity of HPIC-CS-5 column with pyridine-2,6-dicarboxylic
acid eluent.

elutes after these with the CS-2. $Fe^{2+}$, however, is retained longer on the CS-5, eluting after $Cd^{2+}$ and $Mn^{2+}$. It should also be noted from Figure 2 that a new metal, $Hg^{2+}$ is detected. Theoretically, PAR should be capable of detecting many more metals, since it is well known that it reacts with over 20 metals.[4] One of the factors limiting the capability to detect metals in the post-column reaction with PAR is the kinetics of complexation with PAR. For PAR to bind to a metal, the complex between the chelators in the eluent must first dissociate from the metal to be detected. Thus, one of the reasons why only seven transition metal cations were detected in the first studies (see Figure 1) is because the complexes between the chelators and the other metals would not dissociate fast enough. To overcome this problem, certain ligands can be added to the eluent to reduce the time it takes for the complexes with PAR to form. It has been proposed that the added ligands can solvate the metal-chelator complex in such a way as to enhance the metal-PAR complex formation. Ligands which have been found to have such properties include carbonate, ammonia, phosphate, and alkanol amines.[5] The result is that metals which have been previously masked by the presence of eluent chelators, can now be sensitively detected by PAR.

The kinetics of the PAR formation can be outlined as follows:

$$M\text{-(Eluent Chelator)} + \text{Solvating Ligand} \rightarrow M\text{-Ligand} + \text{Chelator}$$

$$M\text{-(Ligand)} + \text{PAR} \rightarrow M\text{-PAR} + \text{Ligand}$$

The solvating ligand rapidly displaces the eluent chelator to form a metal-ligand complex. Even though this complex is formed rapidly, it also dissociates rapidly in the presence of PAR so that the metal-PAR complex can be formed. This technique can be used to analyze more than just standards in deionized water. It has been used to determine transition metals in lake water as shown in Figure 3. Trace concentrations of metals can be determined in the presence of an excess of other metals. This is illustrated in Figure 4 in the determination of lead in beer.

Another factor limiting the post-column reaction with PAR is the eluent pH. In the original work with PAR, the eluent pH was kept relatively high (9.7). This is because PAR, like

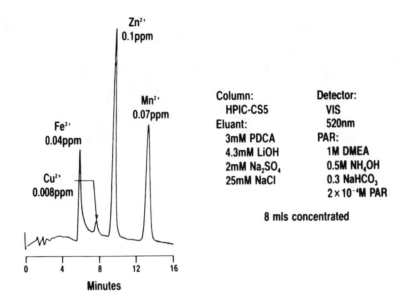

FIGURE 3. Determination of transition metals in natural lakewater. Column: HPIC-CS-5, detection: post-column reaction and visible detector.

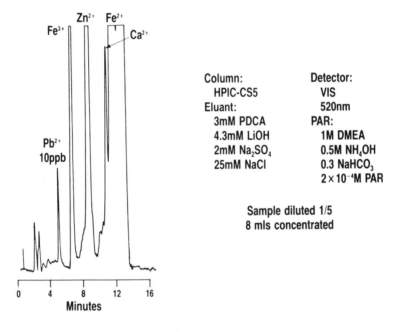

FIGURE 4. Determination of lead in beer. Column: HPIC-CS-5.

most chelators, binds metals better when it is highly ionized. The degree of ionization of PAR increases as the pH increases. However, as the pH increases, so does the amount of hydrolysis of the metals. In other words, more metal hydroxide precipitate appears. The formation of metal hydroxides can be so severe at pH 9.7 that it competes effectively with the formation of metal-PAR complex. By lowering the pH of the PAR reagent, this metal hydrolysis can be limited so that the reaction between the metal and PAR becomes quantitative. As a result, metals that could not be previously detected by PAR can now be

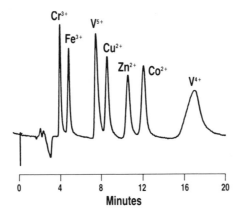

Column:
HPIC-CS5
Eluant:
6mM PDCA
8.6mM LiOH
PAR:
0.5M Na$_2$HPO$_4$
2×10$^{-4}$M PAR
Detector:
VIS
520nm

FIGURE 5.    Determination of vanadium(V) and vanadium(IV). Column: HPIC-CS-5, detection: post-column reaction with PAR; 0.5 *M* sodium phosphate added to enhance kinetics of the post-column reaction.

determined. The scope of metals that can now be determined with PAR post-column reaction is much larger now. One example of this is the determination of $Cr^{3+}$, which is known to form a stable complex with PDCA upon heating. Thus, the sample containing $Cr^{3+}$ is first heated with PDCA to form the Cr-PDCA complex which absorbs at 520 nm. The kinetics of the Cr-PDCA are slow, but ppb-levels can be detected using either a heated post-column reaction coil or by directly monitoring the absorbance at 520 nm and not using PAR. In some cases it is more convenient not to use a heated coil. In these cases, the detection limit for $Cr^{3+}$ is about 1 ppm.[3] The $Cr^{3+}$ is only slightly retained on the CS-5 as shown in Figure 5. It should be noted that 0.5 *M* phosphate was included in the PAR to accelerate the kinetics of metal-PAR formation. Without the PAR, $Cr^{3+}$ could have been detected at lower concentrations, but the $Fe^{3+}$, $Cu^{2+}$, and $Co^{2+}$, would not be detected. It should be added that the determination of $Cr^{3+}$ in the presence of chromic acid, presents special difficulties, and will be discussed in Chapter 4 (Section III).

Vanadium has also been determined using the phosphate ligand and the PDCA eluent. Vanadium is important to steel manufacturers who use it as an additive to make especially strong steel, and to environmentalists who are concerned about its potential toxicity. It is possible to distinguish between the oxidation states of vanadium as shown in Figure 5. Detection limits for vanadium by direct injection were 5 to 10 ppb.[3] Sensitivity to zinc, however, is lowered when the phosphate is added to the PAR. This type of information could be useful when performing qualitative analysis. If an unknown water sample is analyzed without phosphate in the PAR, a large zinc and very small (if any) vanadium peaks would be seen. By re-analyzing the same water sample with phosphate one can perform qualitative analysis. If a peak decreases markedly, and elutes at approximately 10 to 11 min it is probably due to $Zn^{2+}$. If new peaks appear at 8 and 17 min, they may well be due to $V^{5+}$ and $V^{4+}$, respectively.

Gallium can also be determined using the PDCA eluent and by adding phosphate to the PAR. Gallium is becoming increasingly important in the semiconductor industry as GaAs replaces silicon for many applications. Gallium is separated and detected, along with several other potential dopants or contaminants in the GaAs crystal as shown in Figure 6. It is useful to note that the eluent used in this separation and in Figure 5 is different from the one used to separate the ten metals shown in Figure 3. Although the PDCA and LiOH are twice as concentrated in the vanadium and gallium determinations, the eluent used for separating ten metals also contains sodium sulfate and sodium chloride. The sodium elutes the metal cations, based on competition for ion exchange sites on the CS-5. The additional sodium does not

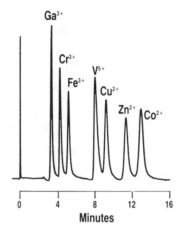

FIGURE 6.    Determination of gallium(III). Column: HPIC-CS-5, detection: post-column reaction with PAR.

appreciably affect the retention of $Fe^{2+}$, but $Cu^{2+}$, $Zn^{2+}$, and $Co^{2+}$ all elute faster in the presence of the additional $Na_2HPO_4$ eluent.

Mercury can also be sensitively determined using the PDCA eluent and by adding phosphate to the PAR. The detection limit for $Hg^{2+}$ using this eluent and phosphate buffer in the PAR is about 0.1 ppm.[3]

Speciation of tin is also possible with the CS-5 column and another eluent. Stannous ion is added to tin-lead alloy plating baths, and can be easily oxidized to Sn(IV). Thus, to monitor plating-bath performance, it is important to know both Sn(II)- and Sn(IV)-levels. The stannous ion, however, presents a unique analytical problem in that it is hydrolyzed at pH >3.[5] As a result, it is important to keep the pH below 3 to ensure complete recovery of Sn(II). As a result, the mixture of PDCA and LiOH used as the eluent in separating other metals (Figure 2) will not work for Sn(II). Fortunately, neither Sn(II) nor Sn(IV) are very strongly retained on the CS-5, and can be eluted with the rather weak eluent, 0.3 m*M* HCl. This eluent adequately separates the two species of tin from the lead in the alloy bath as shown in Figure 7. The ligand found to best accelerate binding of Sn(IV) to PAR, is acetate. Especially in relatively new alloy baths, Sn(IV) is the minor component, requiring that the ion-chromatographic method be optimized for it. As a result, the PAR solution contains 1 *M* sodium acetate, 1 *M* acetic acid, and 0.2 m*M* PAR. A high concentration of ligand is needed to buffer the high acid concentration in the eluent. The combination of eluent plus PAR coming off the post-column reactor produces a pH of about 5.7.

Another series of metals that can be determined with the CS-5 and PAR post-column reaction are the lanthanides. Using an eluent which consists of 2 m*M* pyromellitic acid and 50 m*M* oxalic acid adjusted to pH 3.5 with LiOH flowing at 1 m*ℓ*/min, the lanthanides can be resolved. The PAR reagent contains 3 *M* ammonium hydroxide and 1 *M* acetic acid as ligands, along with 0.05 m*M* PAR. Alternatively, a gradient elution can be used to separate 14 lanthanide metals as shown in Figures 8 and 9. The eluent program was

1.    Eluent 1: Water
2.    Eluent 2: 0.4 *M* 2-hydroxyisobutyric acid, pH 4.6 with LiOH. Gradient: 15 to 100% eluent 2 in 18 min

It should be reemphasized that metals can also be determined as anionic complexes. As discussed in Chapter 1, EDTA complexes with lead, nickel, zinc, and copper can be separated on an AS-5 column.[6] In addition to these, cadmium, manganese, cobalt, and stannous

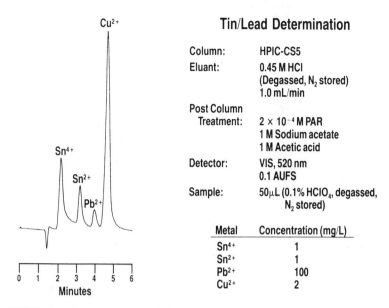

**Tin/Lead Determination**

| | |
|---|---|
| Column: | HPIC-CS5 |
| Eluant: | 0.45 M HCl (Degassed, $N_2$ stored) 1.0 mL/min |
| Post Column Treatment: | $2 \times 10^{-4}$ M PAR 1 M Sodium acetate 1 M Acetic acid |
| Detector: | VIS, 520 nm 0.1 AUFS |
| Sample: | 50 μL (0.1% $HClO_4$, degassed, $N_2$ stored) |

| Metal | Concentration (mg/L) |
|---|---|
| $Sn^{4+}$ | 1 |
| $Sn^{2+}$ | 1 |
| $Pb^{2+}$ | 100 |
| $Cu^{2+}$ | 2 |

FIGURE 7. Determination of tin(II) and tin(IV). Column: CS-5, detection: post-column reaction with PAR. Hydroxylamine added to enhance kinetics of post-column reaction.

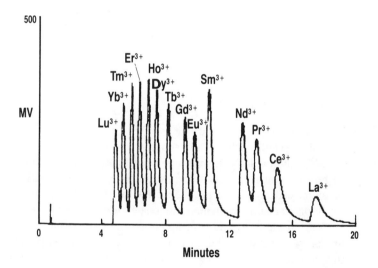

FIGURE 8. Gradient elution ion chromatography of lanthanide metals. Gradient: 85% water/15% 0.4 $M$ 2-hydroxyisobutyric acid, pH 4.6 with LiOH to 100% 0.4 $M$ 2-hydroxybutyric acid.

complexes with EDTA have also been determined by ion chromatography.[7] Divalent metal ions form slightly dissociated complexes with EDTA.[8] EDTA has four carboxylic acid groups, and therefore, four acidic protons. This is often represented in shorthand notation where Y represents EDTA. Depending on the pH, EDTA is represented as $H_4Y$, $H_3Y^-$, $H_2Y^{2-}$, $HY^{3-}$, or $Y^{4-}$. Most metal complexes are formed with the $H_2Y^{2-}$ form. The divalent metal displaces the two acidic hydrogens, i.e.,

$$M^{2+} + H_2Y^{2-} \rightarrow MY^{2-} + 2H^+$$

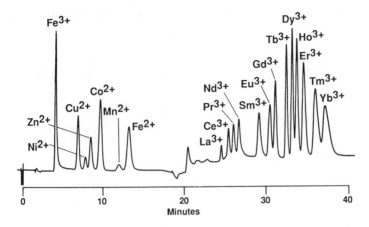

FIGURE 9.   Determination of lanthanide and transition metals in a single injection by ion chromatography. Column: HPIC-CS-5. Gradient: eluent 1: water; eluent 2: 6 m$M$ PDCA; 50 m$M$ sodium acetate, 50 m$M$ acetic acid, pH 4.5; eluent 3: 100 m$M$ oxalic acid, 190 m$M$ LiOH; eluent 4: 100 m$M$ diglycolic acid, 190 m$M$ LiOH; gradient:

| Time | %1 Water | %2 PDCA | %3 Oxalic acid | %4 Diglycolic acid |
|---|---|---|---|---|
| 0 | 0 | 100 | 0 | 0 |
| 12 | 0 | 100 | 0 | 0 |
| 12.1 | 100 | 0 | 0 | 0 |
| 16 | 40 | 0 | 60 | 0 |
| 20 | 40 | 0 | 60 | 0 |
| 20.1 | 20 | 0 | 80 | 0 |
| 28 | 51 | 0 | 26 | 23 |

Thus, the metal-EDTA complex, represented here as $MY^{2-}$, is an anion, and can be separated by anion exchange and detected by chemically suppressed conductivity.

In practice, excess disodium EDTA is added to an unknown solution to ensure that all the metal ions are bound. Thus, there is some free, unbound disodium EDTA. Being a large, hydrophobic anion with a $-2$ charge, the EDTA (or $H_2Y^{2-}$) is strongly retained on anion separators, using a standard bicarbonate/carbonate eluent. It is not retained quite as strongly on the AS-5 as on the AS-1, 2, 3, or 4 columns. By using the AS-5 column, the EDTA can be completely rinsed off faster so that other analyses can be performed. There is one other potential problem. The deionized water used to prepare the bicarbonate/carbonate eluent may contain metal contaminants that would bind the excess, unreacted EDTA in the sample, producing spurious peaks. To prevent this, a metal-free column is placed before the load/inject valve so that metal contaminants are removed from the eluent. As described previously,[6] chloride, nitrate, Pb-EDTA, Ni-EDTA, Zn-EDTA, Cu-EDTA, and sulfate can be separated from each other on an AS-5 column using a 1.5 m$M$ sodium bicarbonate/1.2 m$M$ sodium carbonate eluent. Thus, this method provides simultaneous metal and anion determinations. A metal-free column was placed between the pump and the load/inject valve, to remove metal contaminants in the eluent. In addition, Cd-EDTA, Mn-EDTA, and Co-EDTA can be determined in this way. Not all metal-EDTA complexes can be so determined. The AS-5 column does have a strong affinity for the EDTA in the metal-EDTA complex. If the metal-EDTA complex is not stable enough, the EDTA will dissociate and bind to the AS-5.

Gold, palladium, and platinum complexes with chloride have been determined as

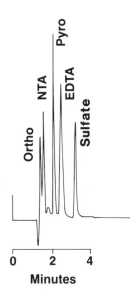

FIGURE 10. Separation of EDTA and sulfate. Column: AS-7, with 0.5 $N$ HNO$_3$ eluent. Detection: post-column reaction with Fe$^{+3}$.

AuCl$_4^-$, PdCl$_4^{2-}$, and PtCl$_6^{2-}$.[9] Au$^{3+}$, Au$^+$, and Co$^{2+}$ have been determined as Au(CN)$_4^-$, Au(CN)$_2^-$, and Co(CN)$_6^{3-}$.[10]

Another type of ion chromatographic analysis that utilizes post-column reaction is the determination of polyvalent anions.[11] Polycarboxylic acids such as nitrilotriacetic acid (NTA), diethylenetetramine pentaacetic acid (DTPA) and ethylenediamine tetraacetic acid (EDTA), along with *ortho-, pyro-,* tripolyphosphate and several polyphosphonates have been shown to react with ferric nitrate (1 g/$\ell$) in 2% perchloric acid to form a colored complex which can be determined by a UV-visible detector set at 430 nm.[11] The separator column used is the AS-7, which has a 10-μm polystyrene/divinylbenzene core particle that is 5% cross-linked. The AS-7 has relatively large latex beads with hydrophobic functional groups (i.e., R in −NR). The AS-7 is especially rugged, being stable to regular use of nitric acid eluents. For example, when using 0.05 $N$ nitric acid, EDTA elutes at approximately 2.5 min and is well separated from sulfate (eluent flow rate in 2 m$\ell$/min) as seen in Figure 10. With a 0.07-$N$ nitric acid eluent, however, sulfate and EDTA co-elute, but tripolyphosphate elutes in a reasonable time as seen in Figure 11. Polyphosphonates used in water treatment require a weaker eluent, 0.03 $N$ nitric acid.[12]

One of the most important applications of post-column reactor detection is the determination of amino acids. Amino acids containing primary amines react with $o$-phthalaldehyde to produce a fluorescent product.[13] Thus, the effluent off the post-column reactor is directed to a fluorescence detector with the excitation wavelength set at 280 nm and the emission wavelength set at 370 nm. If proline, a secondary amine, is being determined, ninhydrin is used for the post-column reaction and the UV-visible detector is set at 570 nm. With ion chromatography, all of the common amino acids can be separated and detected simultaneously in one run on one column without pre-column derivatization. Prior to the ion chromatographic method, it was often necessary to separate the basic amino acids on one column (the short column), and the acidic amino acids on a second column (the long column), and the analysis time could be as long as 80 min.[14] Alternatively, the amino acids could be derivatized with a reagent such as phenylthiohydantoin (PTH). The PTH-amino acid derivatives could then be separated by HPLC with UV detection.[15]

The first ion chromatographic method introduced for amino acid analysis was based on

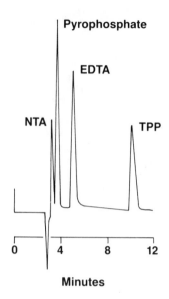

FIGURE 11.    Standard polyvalents. Column: AS-7, detection: post-column reaction with Fe$^{+3}$

ion chromatography exclusion (ICE) followed by electrochemical detection on a pulsed amperometric detector (PAD). This method had the advantage of providing simultaneous determination of carbohydrates and amino acids. Shortly afterwards, a new cation separator, the CS-4, in which the cation-exchange sites were agglomerated latex beads, was specially developed for amino acid analysis. Being zwitterions, amino acids can be converted to their cationic form at low pH. At pH 3.2, even aspartic acid and glutamic acid are cationic and are slightly retained on the CS-4 column, as seen in Figure 12. Histidine, lysine, and arginine are strongly retained with this eluent, so a series of step gradients are needed. If the secondary amine, proline, is also to be determined, the post-column reagent used is ninhydrin as shown in Figure 13. An even more efficient method of amino acid analysis is by anion exchange. An anion separator, the AS-8, was specially designed for amino acid separations. A series of step gradients are needed to separate the amino acids shown in Figure 14. The initial eluent is 36 m$M$ NaOH plus 12 m$M$ sodium borate. In this alkaline eluent, even lysine has a negative charge and is slightly retained. After serine elutes, the eluent is changed to 70 m$M$ NaOH plus 1 m$M$ sodium borate with 2% methanol. After 8 min, the eluent is again switched, this time to a mixture of 175 m$M$ NaOH, 150 m$M$ acetic acid, plus 2% methanol. Finally, after phenylalanine elutes at 12 min, the eluent is switched to 350 m$M$ NaOH, 350 m$M$ acetic acid, plus 2% methanol. Because the AS-8 column has a higher anion-exchange capacity, the 350 m$M$ acetate is retained rather strongly, and should be washed-off before injecting the next amino acid sample. Thus, the AS-8 is regenerated with 300 m$M$ boric acid, 270 m$M$ NaOH, plus 50 m$M$ acetic acid for 3 to 4 min. Fortunately, all these eluent changes can be performed automatically by microprocessor or computer control, permitting completely unattended analysis of amino acid mixtures. The precision of the amino acid analysis is quite good as indicated in Table 2. The percent relative standard deviation, ranges from 0.21 for alanine to 3.20 for tyrosine.

## II. ELECTROCHEMICAL DETECTION

There are several ions which are very weakly ionized and therefore have very low specific conductance. They cannot be detected by chemically suppressed conductivity. Some of these ions will undergo an electrochemical reaction if the appropriate voltage is applied across

## PROTEIN HYDROLYZATE

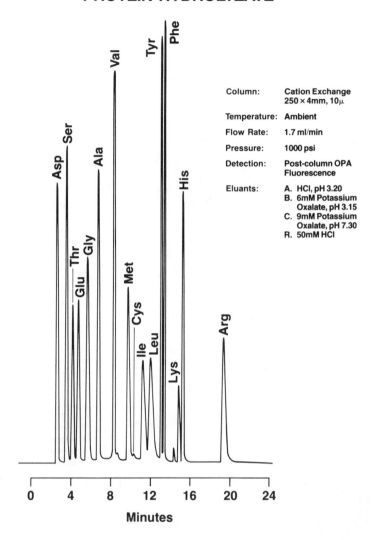

FIGURE 12. Amino acids. Separated as cations. Detection: post-column reaction with *o*-phthalaldehyde and fluorescence.

the proper electrodes. The electrochemical reaction will produce a current which is directly proportional to the concentration of the ion eluting of the separator column. Three of the first ions to be detected in this way were cyanide, sulfide, and bromide. The cyanide and sulfide are weak acids. They can be easily separated from other anions by ion exchange, but like all other anions, they are converted to their acid forms by the suppressor column. Because their pKa values are so high, cyanide and sulfide are in the undissociated HCN and $H_2S$ forms. This is in contrast to HCl, $HNO_3$, and $H_2SO_4$ which have low pKa and exist as the highly conductive pairs of ions, $H^+ + Cl^-$, $H^+ + NO_3^-$, $H^+ + HSO_4^-$. Thus, HCN has low conductivity because it is not dissociated, but HCl has high conductivity because it is dissociated. As a result, cyanide cannot be detected by chemically suppressed conductivity, but $Cl^-$ can.

Fortunately, cyanide is electrochemically active. It will undergo the following reaction at a silver electrode:

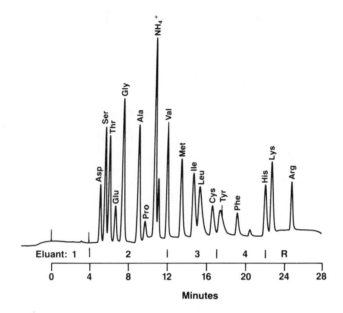

FIGURE 13.    Ninhydrin detection of amino acids, (1 nmol each). Note that the secondary amine, proline, is detected.

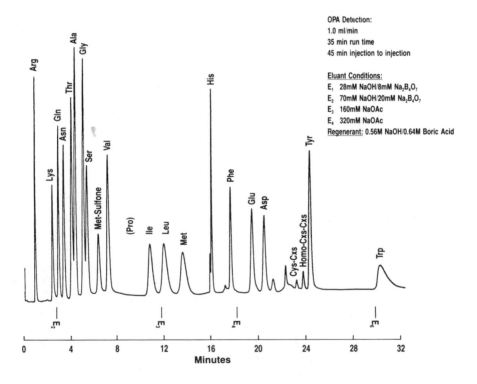

OPA Detection:
1.0 ml/min
35 min run time
45 min injection to injection

Eluant Conditions:
E₁    28mM NaOH/8mM Na₂B₄O₇
E₂    70mM NaOH/20mM Na₂B₄O₇
E₃    160mM NaOAc
E₄    320mM NaOAc
Regenerant: 0.56M NaOH/0.64M Boric Acid

FIGURE 14.    Amino acids by anion exchange.

### Table 2
### AMINO ACID PRECISION
### STUDY, CS-4 CATION
### EXCHANGE, OPA POST-
### COLUMN

| Amino acids | %RSD (6 runs) |
|:---:|:---:|
| Asp | 1.55 |
| Ser | 1.08 |
| Thr | 1.54 |
| Glu | 1.91 |
| Gly | 2.09 |
| Ala | 0.27 |
| Val | 2.32 |
| Met | 0.99 |
| Ile | 2.06 |
| Leu | 1.41 |
| Tyr | 3.20 |
| Phe | 2.04 |
| Lys | 2.03 |
| His | 0.90 |
| Arg | 1.47 |

$$Ag + 2CN^- \rightarrow AG(CN)_2^- + e-$$

This electrochemical reaction was the basis for a flow injection electrochemical analysis for cyanide.[16,17] This work proved that the current produced is directly proportional to the concentration of cyanide, and that the electrode maintains its performance over long periods of time. This is important because many electrochemical reactions have products which poison the electrode surface, thus causing a decrease in sensitivity. The flow injection analysis method suffers from interferences, however, from sulfide and halogens because they also react at the Ag electrode surface, i.e.,

$$2Ag + S^{2-} \rightarrow Ag_2S + 2e-$$

$$Ag + X^- \rightarrow AgX + e-$$

These interferences can be eliminated by separating the ions on an anion-separator column. An amperometric flow-through cell is used as the detector. As described by Rocklin and Johnson,[18] this cell has a 1.3-cm by 0.178-cm silver working electrode, a Ag/AgCl reference electrode, and a platinum counter electrode. A potential is applied between the silver working electrode and the Ag/AgCl reference electrode. The platinum counter electrode allows the potentiostat to maintain a constant applied potential between the working and reference electrodes, and it prevents a damaging current drain on the reference electrode.[18]

The amount of current produced increases until a maximum (a diffusion-controlled plateau) is reached. Above this optimum applied potential, other reactions, such as oxidation of the electrode surface, begin to occur. This would compete with the desired electrochemical reactions. As a result, the current produced falls off. For 1 ppm cyanide, the optimum applied potential is 0.0 V; for 0.5 ppm sulfide it is between −0.1 and 0.0 V; for 5 ppm iodide it is +0.20 V, and for bromide, 0.30 V.[18]

Electrochemical detection can be used in conjunction with chemically suppressed conductivity detection for simultaneous multi-anion analysis. The electrochemical detector is placed between the separator and the suppressor columns of the ion chromatograph. It is important to place the electrochemical detector before the suppressor column. The suppressor lowers the concentration of supporting electrolyte needed for the electrochemical detection. It should be noted that sulfide and cyanide are not seen on the conductivity detector because of their low specific conductances. Bromide, on the other hand, has a relatively high specific conductance and is detected quite readily by chemically suppressed conductivity. This separation was accomplished on an AS-4 column and a 14.7 ppm ethylenediamine, 10 m$M$ sodium borate, and 1 m$M$ sodium carbonate eluent, pH 11.0, flowing at 2.5 m$\ell$/min. It was also demonstrated that 1 ppm cyanide can be accurately determined in the presence of 10 ppm sulfide and 0.04 ppm bromide, or even 1000 ppm chloride. Thus, the bromide content of reagent-grade chloride salts can be determined using the amperometric flow-through cell as a selective detector. If a conductivity detector was used, the chloride peak would swamp the bromide peak.

When determining cyanide in samples containing metal-cyanide complexes, it is important to know when the total cyanide or only the free cyanide is being measured. This depends on the strength of the metal-cyanide complex. Some are weak enough that they are completely dissociated under the chromatographic conditions being used. This can be thought of as a competition between the anion exchange column and the metal for cyanide ligand. The strength of the metal-cyanide complex is measured by the combined dissociation constant, $\beta$. $Cd(CN)_4^{2-}$ (log $\beta$ = 18.78) and $Zn(CN)_4^{2-}$ (log $\beta$ = 16.7) are labile enough, that all the cyanide dissociates from the metal and elutes as free cyanide. Thus, both free and bound cyanide are detected in the presence of cadmium and zinc. $Ni(CN)_4^{2-}$ (log $\beta$ = 31.3) and $Cu(CN)_4^{2-}$ (log $\beta$ = 30.3) are less labile. They are retained on the AS-4 column and slowly dissociate, producing peak tailing at the electrochemical detector.[19] Still others, such as $Au(CN)_2^-$ (log $\beta$ = 38.3), $Fe(CN)_6^{3-}$ (log $\beta$ = 42), and $Co(CN)_6^{4-}$ (log $\beta$ = 64), are not at all labile, and no free cyanide can be detected when these complexes are injected on the AS-4. To determine these three metal-cyanide complexes, MPIC with chemically suppressed conductivity detection, is used (see Chapter 3 [Section III]).

It should be noted, that in the paper by Rocklin and Johnson,[18] the AS-4 was used to separate cyanide and sulfide. For iodide determination, however, the AS-1 column with 20 m$M$ sodium nitrate eluent was used. For bromide, the AS-3 with 2 m$M$ sodium carbonate eluent was used. In still another report,[20] sulfide and cyanide were separated on an AS-1 column using a NaOH eluent. Other workers[21] put the electrochemical detector after a short guard column and used the standard bicarbonate/carbonate eluent. Bromide, iodide, and thiocyanate eluted at 1.5, 6, and 9 min. The effluent was then directed to a full-sized separator column and onto a suppressor column so that simultaneous electrochemical and conductivity detection could be used. They reported the detection of chloride, bromide, iodide, cyanide, thiocyanate, dithionate, and sulfide, using a silver electrode with an applied potential of +0.2 V. In a different report,[22] linear calibration plots for cyanide were obtained down to the ppb range. An interesting application of electrochemical detection was also reported by Tarter.[23] He reported the determination of nonoxidizable species using a silver electrode and an applied potential of +0.30 V. An AS-3 column with bicarbonate/carbonate eluent was used. The electrochemical detector was placed just past the anion fiber suppressor. In this way, the electrochemical detector became sensitive to changes in pH. He reported the detection of fluoride, chloride, phosphate, nitrate, and sulfate.

More recently, a pulsed amperometric detector has been developed for the electrochemical detection of −CHOH-containing compounds. Instead of applying one fixed potential to the working electrode, three different voltages are applied in distinct pulses. In the case of carbohydrates, a −CHOH group is oxidized to a C=O which collects at the gold electrode. The C=O is then cleaned off the surface of the gold electrode with a second pulse at a higher voltage. Because the gold electrode itself is partially oxidized to gold oxide, it is cleaned by applying a large negative voltage. This cycle of pulses repeats itself every 0.36 sec.

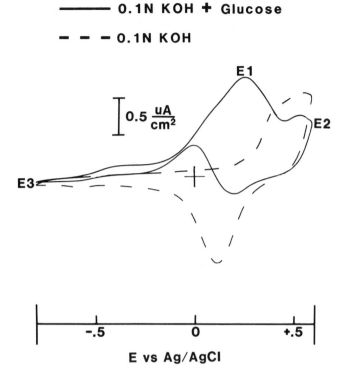

FIGURE 15. Cyclic voltammetry of glucose on a gold electrode. 0.1 *N* KOH is the eluent used to separate carbohydrates.

This technique of pulsed amperometric detection is especially useful for the determination of sugars and alcohols which have a very low specific conductance and are poor UV absorbers. Prior to the introduction of the pulsed amperometric detector, refractive-index detectors were used. They have low sensitivity and can be difficult to use. Carbohydrates can also be determined by gas chromatography, but only after time-consuming derivatization to form a volatile compound.[24]

The choice of voltages used in the detection of carbohydrates is based on cyclic voltammetry data. The applied voltage is continuously varied and the current is monitored. A cyclic voltamagram for the eluent, 0.1 *N* NaOH, and for glucose in 0.1 *N* NaOH is shown in Figure 15. The dashed line shows the background scan in the presence of the 0.1 *N* NaOH supporting electrolyte which is used as the eluent in the ion chromatographic separation of carbohydrates. It shows the oxidation of gold beginning at 0.25 V (the large positive peak which reaches a maximum at +0.5 V). The gold oxide is reduced back to gold at about 0.1 V as indicated by the large negative (cathodic) peak. When glucose is added (solid line), oxidation begins at −0.5 V on the positive-going scan, and produces a large positive (anodic) peak at 0.26 V. This is due to the oxidaton of −CHOH to −C=O. On the negative-going scan, the gold oxide is reduced, producing a small negative (cathodic) peak at about 0.1 V, followed by a larger positive (anodic) peak at about 0.0 V. This implies that the oxidation of glucose which occurs at the large gold surface is inhibited by the formation of gold oxide. As soon as the gold oxide is reduced to free-gold, the glucose which had diffused towards the gold surface is oxidized, producing the positive peak at 0.0 V.

If a single potential is applied to the gold electrode, glucose can be easily detected the first time, but the signal decreases as more samples are detected. The products from the oxidation of glucose deactivate the surface of the electrode. Thus, it is necessary to remove

## PULSED AMPEROMETRIC DETECTOR
## APPLIED POTENTIAL SEQUENCE

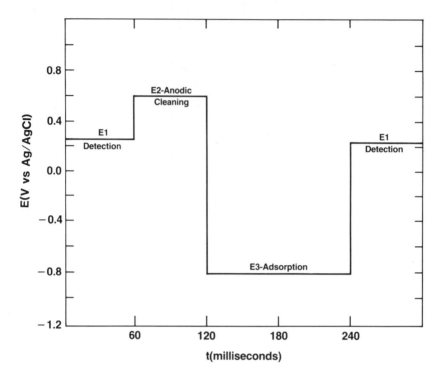

FIGURE 16.    Pulsed sequence for CHOH groups.

these products by pulsing the electrode from $-0.8$ to $0.2$ V, and measure the oxidation at $0.2$ V. The $-0.8$-V pulse reduces the gold oxide to free-gold. If a positive cleaning potential is also included in the sequence of pulses, reproducibility is optimized. The sequence of pulses is summarized in Figure 16.

To separate the carbohydrates, a unique column and separation scheme is used. Previous attempts at liquid chromatography of carbohydrates utilize high-capacity cation exchange columns in the metal form,[26-28] or microparticulate silica with a chemically bonded amino phase,[29] or separation as borate complexes.[30] This new ion chromatographic method takes advantage of the fact that carbohydrates are anions at very high pH (their pKa values range from 12 to 14) and utilize a new anion separator, the AS-6. The AS-6 column has a higher capacity than the AS-4A or AS-5. To make the carbohydrates anionic, sodium hydroxide concentrations as high as $0.1$ $N$ are needed. The AS-4A and AS-5 columns would be overloaded with this concentrated eluent. Because of its higher capacity, the AS-6 is not overloaded. In addition, alcohols will react with the platinum electrode in the pulsed amperometric detector as shown in Figure 17. They are separated on an ICE column.

Sugar alcohols and saccharides were separated on an AS-6 using $0.15$ $N$ NaOH eluent at $36°C$.[25,31] Sugars in soy flour can be separated as shown in Figure 18. If a more dilute eluent is used, smaller sugars can be resolved as shown in Figure 19. This has an important application in the determination of common monosaccharides found in plant hydrolysates, as shown in Figure 20. The sample was prepared by diluting it 1:100 in water, then by passing the diluted sample through a reversed-phase sample-pretreatment cartridge.

Disaccharides and oligosaccharides are more strongly retained and require a more concentrated eluent. The eluent strength can be increased by adding sodium acetate and more

1    **Glycerol 2.5 PPM**
2    **Ethylene Glycol 2.5 PPM**
3    **Methanol 1.75 PPM**
4    **Ethanol 10 PPM**
5    **2-Propanol 15 PPM**

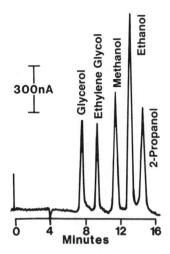

FIGURE 17.    Determination of alcohols. Column: ICE, detector: pulsed amperometric.

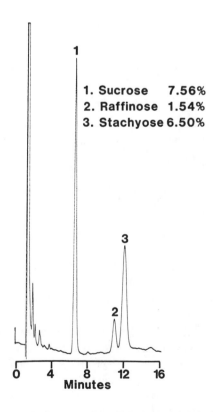

1. Sucrose    7.56%
2. Raffinose    1.54%
3. Stachyose 6.50%

FIGURE 18.    Sugars in soy flour. Pulsed amperometric detection.

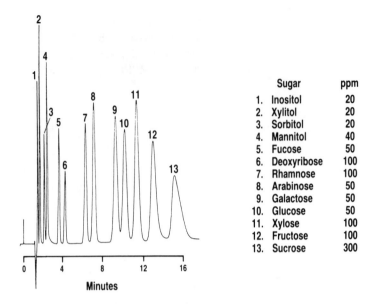

| Sugar | ppm |
|-------|-----|
| 1. Inositol | 20 |
| 2. Xylitol | 20 |
| 3. Sorbitol | 20 |
| 4. Mannitol | 40 |
| 5. Fucose | 50 |
| 6. Deoxyribose | 100 |
| 7. Rhamnose | 100 |
| 8. Arabinose | 50 |
| 9. Galactose | 50 |
| 10. Glucose | 50 |
| 11. Xylose | 100 |
| 12. Fructose | 100 |
| 13. Sucrose | 300 |

FIGURE 19.   Anion exchange separation of carbohydrates.

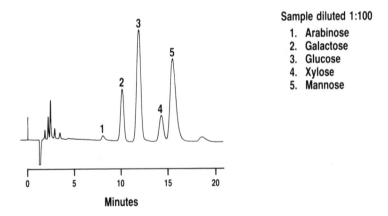

Sample diluted 1:100

1. Arabinose
2. Galactose
3. Glucose
4. Xylose
5. Mannose

FIGURE 20.   Monosaccharides in hydrolyzed wood pulp. Pulsed amperometric detection.

NaOH. Thus, oligosaccharides with degrees of polymerization 2 to 9 were separated using 0.2 *M* NaOH plus 0.2 *M* sodium acetate at 34°C, as shown in Figure 21. There are a few factors which can be used to control selectivity and retention. First, as the NaOH concentration increases, not only does the capacity factor (k′) for carbohydrates and oligosaccharides decrease, but also column efficiency increases. For glucose, the efficiency doubles when going from 0.1 to 1.0 *M* NaOH. Another factor is column temperature. As the temperature increases, retention (i.e., k′) decreases. This temperature effect is larger for oligosaccharides than for disaccharides or monocarbohydrates. At the upper temperature limit, tailing and secondary peaks are observed for some carbohydrates. Column efficiency decreases above 45°C and the anion exchange materials are degraded.[31]

The detection limits for carbohydrates range from about 30 ppb for sugar alcohols to 100 ppb for oligosaccharides. Plots of peak height vs. concentration are linear up to approximately 1000 ppm, at which point the capacity of the AS-6 is overloaded.

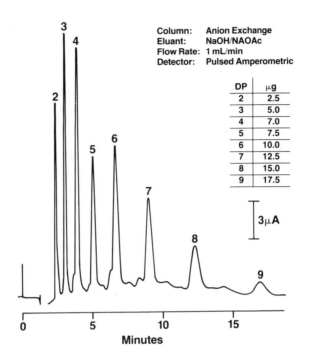

FIGURE 21. Oligosaccharides of maltose. Degrees of polymerization (DP) 2 to 9.

Thus, an electrochemical detector can be used in the determination of a variety of substances. Cyanide, sulfide, phenols, nitrobenzenes, hypochlorite, bromide, and iodide can be detected using a single applied potential. Alcohols and carbohydrates, on the other hand require a pulsed amperometric detector, where three different potentials are used.

## III. MOBILE-PHASE ION CHROMATOGRAPHY (MPIC)

Some anions, such as metal cyanides, alkyl sulfonates, aryl sulfonates, and substituted benzoic acids, are very strongly retained on anion separator columns and require a completely different technique for separation. To satisfy this need, mobile-phase ion chromatography (MPIC) was developed. The MPIC column is packed with a microporous polystyrene/divinylbenzene resin. There are no sulfonate or quaternary amine groups attached, so there is no permanent, fixed charge. The polystyrene/divinylbenzene is synthesized by copolymerization of styrene and divinylbenzene in the presence of poragens, such as volatile organic compounds. After polymerization is complete, any volatile organics can be removed by heating in a vacuum. The resulting resin has a higher surface area than the pellicular columns described earlier. Large hydrophobic analytes are therefore exposed to large surface areas so that there is significant interaction with the nonionic polystyrene/divinylbenzene. The mechanism for retention is by nonpolar interactions such as van der Waals forces. This is analogous to the well-known technique of reversed-phase HPLC. The major differences between reversed-phase HPLC and MPIC are

1. The MPIC column is packed with polystyrene/divinylbenzene which is stable from pH 0 to 14.
2. Chemically suppressed conductivity is very often used with MPIC but not reversed-phase HPLC (which usually utilizes a UV-visible detector).

It is well known that anions (and cations) are not well retained on nonionic columns. To increase retention and enhance resolution, the anions are converted to neutral ion pairs by adding a suitable ion-pair reagent. In other words, a quaternary ammonium compound such as tetrapropyl ammonium hydroxide (TPAOH) will form an uncharged ion pair with an anion such as thiocyanate. The ion pair in this case can be designated as $TPA^+SCN^-$. This ion pair is retained on the MPIC column, where $SCN^-$ alone, is not. After the $TPA^+SCN^-$ elutes off the column, it is detected by chemically suppressed conductivity.

The key to a successful separation of large, hydrophobic anions is the proper selection of eluent. Several factors will control the strength of the eluent. They include the type and concentration of ion-pairing reagent, the type and concentration of organic solvent added, the presence of inorganic modifiers, pH, and temperature. The most common ion-pair reagents for anion analysis are ammonium hydroxide, tetrapropyl ammonium hydroxide (TPAOH), and tetrabutyl ammonium hydroxide (TBAOH). One model for explaining the retention mechanism is the formaton of uncharged ion pairs. According to this model, an equilibrium is established between the ion-pair reagent and the analyte anion. In the case where TPAOH, or $(C_3H_7)_4N^+$, the equilibrium is as follows:

$$(C_3H_7)_4N^+ + A^- \rightarrow (C_3H_7)_4N^+A^-$$

(ion pair)

As the concentration of ion pair increases the amount of ion pairs also increases. After a maximum amount of ion-pair reagent (TPAOH in this example) is added, all the analyte anions ($A^-$) are tied up as ion pairs, and no further increase in ion-pair reagent concentration will affect analyte retention.

The choice of ion-pair reagent depends on the hydrophobicity of the analyte anion. If the analytes are hydrophilic, i.e., fluoride, chloride, nitrite, bromide, and nitrate, the very hydrophobic TBAOH may be needed to form an ion pair which is hydrophobic enough to bind to the polystyrene/divinylbenzene in the MPIC column. As described previously,[32] if ammonium hydroxide was used as the ion-pair reagent, none of the hydrophilic anions are retained on the column. If the TPAOH is used, retention is observed, but nitrite and bromide are not well resolved. When TBAOH is used, the retention is so strong that fluoride elutes in about 13 min. To elute the other $TBA^+A^-$ ion pairs, an organic solvent such as acetonitrile must be added to the eluent. With 8% acetonitrile, separation is rather weak.

If the analyte anions are hydrophobic, a hydrophilic ion-pairing reagent such as ammonium hydroxide is needed. For example, alkyl sulfonate detergents such as pentyl and hexyl (C-5 and C-6) sulfonates elute at about 5 and 23 min when ammonium hydroxide is used as the ion-pair reagent. To shorten the elution time, and enable the elution of C-7 sulfonate, 5% acetonitrile is needed as shown in Figure 22. If the acetonitrile concentration is increased to 10%, C-8 sulfonate can also be eluted, but C-5 and C-6 are eluted so quickly that they are not well separated. If TPAOH were used as the ion-pair reagent, more acetonitrile (20%) is needed to elute C-8 in a reasonable time. If TBAOH is used, 28% acetonitrile is needed to elute C-8. Thus, different combinations of acetonitrile (or even methanol) concentration and selection of ion-pair reagent, can produce similar MPIC separations. In general, the retention of anions increases with the hydrophobicity of the ion-pair reagent, i.e., $NH_4OH$ < TPAOH < TBAOH. On the other hand, retention decreases as acetonitrile is added to the eluent. It is possible, however, to elute very hydrophobic detergents such as dodecyl-sulfate and tetradecylsulfate as shown in Figure 23.

Similarly, aromatic anions, such as aromatic sulfonates can be separated by MPIC. As shown previously,[32] aromatic sulfonates elute much earlier than their corresponding alkyl sulfonates. For example, using an eluent that elutes benzene sulfonate in approximately 2.5

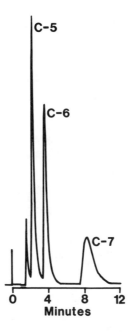

FIGURE 22.   Separation of linear alkyl sulfonates. Column: MPIC; detection: conductivity.

**Eluent: 10mM NH₄OH
in 32% Acetonitrile**

**Detector Sensitivity: 10μS**

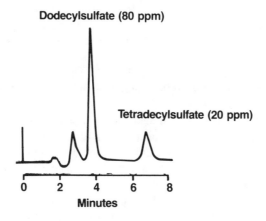

FIGURE 23.   Analysis of hydrophobic anions by MPIC. Anionic surfactants.

min, the alkyl (C-6) sulfonate would elute in approximately 23 min. Similarly, substituted benzoic acids can be separated by MPIC. In one application, liquid laundry detergent is analyzed for xylene sulfonate.

When selecting the proper ion-pair reagent, a general guideline is that for separating very similar compounds, a smaller ion-pair reagent will provide better resolution. Because the ion-pairing reagent is so small, the dominant portion of the structure of the ion pair will be the compound being separated. The separation is achieved more by the molecular structure of the analyte, than the ion-pairing reagent. If the relatively large TBAOH was used as the

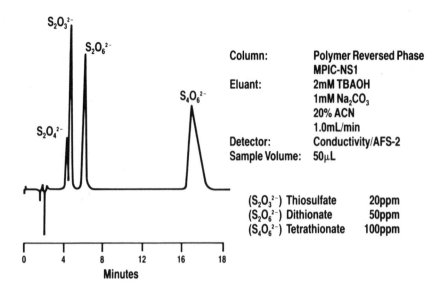

FIGURE 24.    Determinations of larger sulfur oxides. Column: MPIC.

ion-pairing reagent, the structural and chemical properties of the ion pair would be dominated by the TBAOH, and resolution would not be so good. There are some anions, however, which are simply not retained on the MPIC column unless paired with the large, hydrophobic TBAOH. The common hydrophilic anions are a good example. Even though the structure of the $TBA^+F^-$, $TBA^+Cl^-$, $TBA^+NO_2^-$, $TBA^+Br^-$, and $TBA^+NO_3^-$ are dominated by the $TBA^+$ portion, resolution is still quite good. On the other hand, dodecylsulfate and tetradecylsulfate are very similar anions and have a strong affinity for the MPIC column the only possibility for eluting them is to use ammonium hydroxide as the ion-pairing reagent. Using ammonium hydroxide and 32% acetonitrile, they are quite well separated as shown in Figure 23. Thus, when selecting an ion-pairing reagent it is necessary to consider the relative hydrophobicity of the anions and whether their chemical structures are similar or different. If the anions are rather hydrophilic, a larger ion-pairing reagent (TBAOH) is indicated. If the anions have very similar chemical structures, a smaller ion-pairing reagent (ammonium hydroxide), is used.

Another factor that can affect the strength of the eluent in MPIC is the addition of an inorganic modifier such as sodium carbonate for anion analysis. The retention of divalent anions such as sulfate and phosphate is affected more by sodium carbonate than are monovalent anions. Using 8% acetonitrile with TBAOH, monovalent anions are eluted within 10 min, but sulfate and phosphate are eluted within 10 min without sacrificing resolution of the monovalent anions. The amount of sodium carbonate that is needed depends on what other analyte anions are present in the sample, but usually a concentration of 0.1 to 1.0 m$M$ sodium carbonate is used. When 1 m$M$ sodium carbonate is included in an eluent consisting of 20% acetonitrile and 2 m$M$ TBAOH, the large, hydrophobic oxy acids of sulfur, thiosulfate, dithionate, and tetrathionate can be separated as shown in Figure 24. Similarly, other hydrophilic anions can also be separated as shown in Figure 25.

When determining anions, it may be necessary to lower the eluent pH. It is better to do this by adding boric acid which has very low conductivity compared to HCl, $HNO_3$, or $H_2SO_4$. Lowering the pH may be necessary if the analyte anions are unstable at a higher pH or if it is necessary to lower the charge of an anion (i.e., convert $HPO_4^{2-}$ to $H_2PO_4^-$). For example, some mercaptans form disulfides at high pH and some phospholipids are hydrolyzed at high pH. In these cases, it may be necessary to adjust the eluent pH to 7, by adding boric acid.

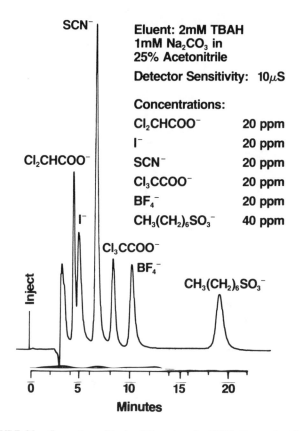

FIGURE 25.   Separation of hydrophilic anions by MPIC. Inorganic anions.

   With the advent of gradient elution ion chromatography, it is now possible to gradually alter the strength of the eluent by increasing the acetonitrile concentration. The background conductivity will remain rather constant, especially if the concentrations of ion-pairing reagent and inorganic modifier are not changed. Thus it is possible to begin an analysis with low acetonitrile concentration to achieve separation of the more hydrophilic anions which co-elute at higher acetonitrile concentrations. By increasing the acetonitrile concentration, the more hydrophobic anions can then be eluted.

   There is another proposed mechanism for ion-pair chromatography which is supported by experimental evidence. This retention model, proposed by Iskandorani and Pietrzyk,[33] introduces the concept of an electric double layer. The boundary between the nonpolar polystyrene/divinylbenzene and the polar aqueous eluent is similar to the boundary between octane and water. It is well known that if a detergent is present, it will line up between the boundary with its hydrophobic tail extending into the nonpolar octane and its hydrophobic head extending into the polar water phase. Thus, the $CH_2CH_2CH_2CH_2CH_2CH_2$ of the hexane sulfonic acid (used in cation analysis) would extend into the octane layer and the $-SO_3H$ would extend into the water layer. Similarly, the hexyl tail would extend into the nonpolar polystyrene/divinylbenzene of the MPIC column and the $-SO_3H$ would extend into the aqueous eluent. Because $-SO_3H$ is a strong acid, the $H^+$ is dissociated and can migrate a small distance from the $-SO_3^-$. Thus, an electropotential is established between the plane of the ionic $-SO_3^-$ and the eluent. The potential will be dependent on the amount of detergent absorbed onto the polystyrene/divinylbenzene. The $H^+$ ions are considered to be distributed into two different zones. Some of the $H^+$ ions are located very close to the $-SO_3^-$, in a

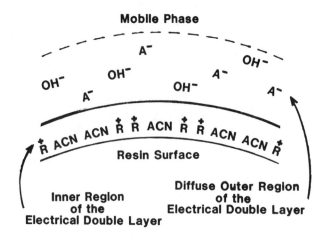

FIGURE 26.    Schematic representation of the electric double layer in MPIC. ACN, acetonitrile.

zone called the Stern layer, which has a depth on the order of $10^{-8}$.[34] The remaining $H^+$ ions are located further from the $-SO_3^-$ in a diffuse region called the Gouy layer.

For anion analysis, a tetraammonium hydroxide such as TPAOH would be used as the ion-pairing reagent. In this case, the propyl groups would extend into the polystyrene/divinylbenzene and the $-N$ would extend into the aqueous eluent. The $OH^-$ and any other inorganic modifier (i.e., $CO_3^{2-}$ from $Na_2CO_3$) would form the Stern and Gouy layers. Any acetonitrile in the eluent would compete for binding sites on the polystyrene/divinylbenzene. Thus, an electric double layer similar to that in Figure 26 would be established.

Iskandorani and Pietrzyk[33] described the equilibria that would be established in such a system. These equilibria were written in general terms, accounting for any possible combination of ion-pairing reagent and inorganic modifier (i.e., counter ion). These equilibria can be rewritten using the specific example of a tetraammonium hydroxide, $R_4N^+$ as the ion-pairing reagent and $OH^-$ as the counter anion. Polystyrene/divinylbenzene is the stationary phase, and is abbreviated to PS/DVB.

$$K1$$
$$R_4N^+ + OH^- + PS/DVB \rightarrow PS/DVB-R_4NOH$$

$$K1 = \frac{(PS/DVB-R_4NOH)}{(R_4N^+)(OH^-)(PS/DVB)} \tag{1}$$

Next, consider the binding of the ion pair consisting of $R_4N^+$ bound to the analyte anion, $X^-$, i.e.,

$$K2$$
$$R_4N^+ + X^- + PS/DVB \rightarrow PS/DVB-R_4NX$$

$$K2 = \frac{(PS/DVB-R_4NX)}{(R_4N^+)(X^-)(PS/DVB)} \tag{2}$$

Third, consider an ion exchange where $X^-$ exchanges with OH– on the column, i.e.,

$$\text{PS/DVB–R}_4\text{NOH} + \text{X}^- \xrightarrow{\text{K3}} \text{PS/DVB–R}_4\text{NX} + \text{OH}^-$$

$$\text{K3} = \frac{(\text{PS/DVB–R}_4\text{NX})(\text{OH}^-)}{(\text{PS/DVB–R}_4\text{NOH})(\text{X}^-)} \tag{3}$$

The concentration of active binding sites on the polystyrene/divinylbenzene, designated as (PS/DVB) in the equations above, can be thought of as the sorption capacity, or the moles of detergent (ion-pairing reagent) sorbed per gram of polystyrene/divinylbenzene. The sorption capacity, K, can be represented by:

$$\text{K} = (\text{PS/DVB}) + (\text{PS/DVB–R}_4\text{NOH}) + (\text{PS/DVB–R}_4\text{NX}) \tag{4}$$

Equations 1 to 4 can be combined with elimination of (PS/DVB) and (PS/DVB–R$_4$NOH) and solving for (PS/DVB–R$_4$NX), i.e.,

$$(\text{PS/DVB–R}_4\text{NX}) = \frac{\text{K}(\text{OH}^-)}{(1/\text{K2})(\text{R}_4\text{N}^+) + \text{K1}(\text{OH}^-)/\text{K2} + (1 + \text{K3K1/K2})(\text{OH}^-)} \tag{5}$$

Retention of the analyte, X$^-$, is given by:

$$\text{k}' = \text{q}(\text{PS/DVB–R}_4\text{NX})/(\text{X}^-) \tag{6}$$

where k$'$ is the capacity factor and q is the ratio of stationary-phase volume to mobile-phase volume.

Equations 5 and 6 can be combined to give:

$$\frac{1}{\text{k}'} = \frac{1}{(\text{q})(\text{K0})(\text{K1})(\text{R}_4\text{N}^+)} + \frac{\text{K1}(\text{OH}^-)}{(\text{q})(\text{K0})(\text{K1})} + \frac{1}{(\text{q})(\text{K0})}\left(1 + \frac{(\text{K3})(\text{K1})}{\text{K2}}\right.$$

which can be rewritten as:

$$\frac{1}{\text{k}'} = \frac{1}{\text{A}(\text{R}_4\text{N}^+)} + \text{B}(\text{OH}^-) + \text{C}(\text{X}^-) \tag{7}$$

This equation predicts that the retention time, or capacity factor, of an analyte anion, k$'$, is directly proportional to the concentration of the ion-pairing cation, R$_4$N$^+$, and indirectly proportional to the concentration of the counter anion, OH$^-$, and indirectly proportional to the concentration of the analyte itself. In practice, it is impossible to increase the concentration of the ion-pairing cation, R$_4$N$^+$, by itself. In the example used to derive these equations, R$_4$N$^+$ is added as the hydroxide, i.e., TPAOH. In fact, the retention times of anions increase as the concentration of ion-pairing reagent increases until a maximum is reached.[33] Above a certain maximum concentration of ion-pairing reagent, the capacity factor begins to decrease. This can be interpreted in two different ways. First, it is possible that as more ion-pairing reagent is added, more of the analyte anion is bound as an ion pair. At some point, all the anion is bound, and no further increase in ion-pairing reagent can bind any more anion. At this point, the system is saturated with ion-pairing reagent, and the retention time of the analyte will not increase further. On the other hand, it is possible that since the addition of more ion-pairing cation, $R_4N^+$, is necessarily accompanied by an addition of more counter anion, $OH^-$. The two exert opposite influences, $R_4N^+$ increases the retention time, and OH− tends to decrease the retention time of the analyte anions.

It is quite possible to use additional counter anions other than OH−. Iskandaroni and Pietrzyk[33] used several different counter anions. They found that, in general, for most analytes the particular counter anion used was important. The strength of the counter anion in reducing capacity factor (retention time) was $Br^- > NO_3^- > Cl^- >$ citrate $>$ formate $>$ phosphate $> SO_4^{2-} > F^- > OH^-$.

When using chemically suppressed conductivity detection, however, it is important to use ion-pairing reagents with a minimum of ionic impurities. This encourages the use of hydroxide as the counter anion since ammonium hydroxide, TPAOH, and TBAOH can be obtained in relatively pure form. Thus, if it is useful to decrease the retention times of analytes, counter anions can be added as sodium salts, i.e., NaBr, NaCl, $Na_2SO_4$, or $Na_2CO_3$.

Of course, there is no reason why cations cannot be determined by MPIC as well. In an acidic eluent, aliphatic and aromatic amines are cations. The eluent can be made acidic by use of an alkyl sulfonate such as hexane or octane sulfonic acid, which can form an ion pair with an amine. In the determination of monoethanolamine, octane sulfonic acid is used and the following equilibrium is established:

$$CH_3(CH_2)_6CH_2-SO_3 + H_2NCH_2CH_2OH$$

$$\rightarrow CH_3(CH_2)_6CH_2SO_3-H_2NCH_2CH_2OH$$

When determining ppm levels of monoethanolamine, 5 m$M$ octane sulfonic acid is sufficient to fully protonate the amine and tie it all up as an ion pair. Detection is achieved using chemically suppressed conductivity. In the original reference,[35] it is suggested that a packed-bed suppressor be used and be converted to its borate form. The reason why the packed-bed suppressor was originally recommended is because no other suppressor column was available at the time. Since then, the cation micromembrane suppressor was developed. Unlike the packed-bed suppressor, the cation micromembrane suppressor (CMMS) is continually regenerated, permitting automation. Both the packed-bed suppressor and the CMMS are resistant to the acetonitrile and octane sulfonic acid in the eluent. Thus, the separation of mono-, di-, and triethanolamine shown in Figure 27 can now be conveniently performed using a CMMS. In a similar application, the organic cation cetylpyridinium can be determined in commercial mouthwash as shown in Figure 28.

The electric double layer model for anion separations is essentially the same as for cation separations. Again, the nonpolar polystyrene/divinylbenzene and the polar, aqueous eluent can now be thought of as two separate phases similar to octane and water. The hydrophobic tail of the sulfonate detergent is located in the nonpolar polystyrene/divinylbenzene phase and the hydrophobic $-SO_3^-$ head of the detergent extends into the aqueous eluent. The $H^+$ counter ions are located in two different regions. Some of them are very close to the $-SO_3$ groups in the Stern layer. The remaining $H^+$ is located further from the $-SO_3$ in the diffuse Gouy layer. Equilibria similar to those described for anions (in Equations 1 to 6) exist in cation separations.

Instead of using $-NR_3$ as the ion-pairing reagent, $-RSO_3$ is used. Instead of $OH^-$ being the counter ion, $H^+$ is the counter ion for cation separations, and the cation analytes would be generalized as $X^-$. An equation can be derived which predicts that the retention time, or capacity factor, will increase as the concentration of ion-pairing reagent increases. The capacity factor is predicted to decrease as the concentration of the counter ion, $H^+$, and analyte, $X^-$, increase.

MPIC can be used to separate aliphatic and aromatic amines. Aromatic amines can be

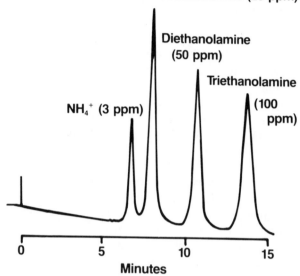

FIGURE 27. Analysis of hydrophilic cations by MPIC. Ethanolamines. Eluent: 2 m*M* hexanesulfonic acid (HSA) in 100% water.

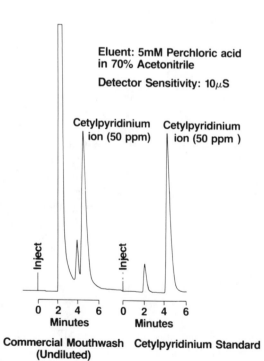

FIGURE 28. Analysis of hydrophobic cations by MPIC: quaternary germicides.

detected by a UV-visible detector, and, if their specific conductance is high enough, by chemically suppressed conductivity.

There is no reason to use only chemically suppressed conductivity for detection. MPIC was originally developed for the determination of large hydrophobic anions, and aliphatic amines which are strongly retained on ion-exchange separator columns. There are, of course, a number of other ions which have other easily measured properties. Aromatic ions might be better detected by a UV-visible detector. Benzoic acids, being anions at high pH, can form ion pairs with either TPAOH or TBAOH and be separated by the same mechanism as the inorganic ions described. In addition, there might be additional interaction between the polystyrene/divinylbenzene resin and the aromatic groups on substituted benzoic acids. Similarly, aromatic amines are cations at low pH and can form ion pairs with sulfonic acids. Again, there also might be interactions between the aromatic part of the amines and the polystyrene/divinylbenzene. In some cases, a benzoic acid or aromatic amine might have a large enough specific conductance to be detected with better sensitivity by a conductivity detector instead of a UV detector. Other benzoic acids (e.g., 3-fluorobenzoic acid) and aromatic amines (e.g., *m*-toluidine) which have low specific conductances, but appreciable absorptivity, can be better detected using a UV detector. Thus, it is possible to simultaneously determine aliphatic and aromatic amines by using both a UV and a conductivity detector.

Another useful detection mode for MPIC is electrochemical detection. When an appropriate voltage is applied to a glassy carbon electrode, certain organic compounds are oxidized and a current is produced. Phenols, catecholamines, aromatic amines, and nitro compounds are all capable of being oxidized if the proper voltage is applied to a glassy carbon electrode. Nitrobenzenes can be determined by MPIC with electrochemical detection. An eluent consisting of 50% aqueous $1.0\,M$ lithium perchlorate, $0.2\,M$ perchloric acid, and 50% acetonitrile will separate nitrobenzene and benzene.[36] They are detected by applying $-0.80$ V to the glassy carbon electrode.[36] Phenols can be separated using an eluent consisting of 60% aqueous $0.2\,M$ sodium phosphate (monobasic) and 40% acetonitrile as shown in Figure 29. They are detected by applying 1.25 V to either a glassy carbon or platinum electrode.

The example of nitrobenzenes and phenols illustrates another important point. In the eluents used, neither the phenols nor the nitrobenzenes were ionic. The mechanism for their retention must be nonionic interactions such as van der Waals forces. This is quite analogous to reversed-phase HPLC using chemically bonded octadecyl (C-18) silica columns. Though the chromatographic properties of polystyrene/divinylbenzene have not been thoroughly investigated, there are some important distinctions. First, the polystyrene/divinylbenzene is pH resistant. This is especially important when analyzing highly caustic or acidic samples. To analyze such corrosive samples with a silica-based column, it is necessary to neutralize the sample. Second, the polystyrene/divinylbenzene resin has a large phenyl content, permitting aromatic-aromatic interactions which are known to be an important force in organics containing phenyl groups.[37] Third, the polystyrene/vinylbenzene does not have the same physical properties as a C-18 column. The C-18 (octadecyl) has much more mobility than the rigid polystyrene/divinylbenzene. The C-18 group can almost be thought of as a liquid. In reversed-phase HPLC using a C-18 column, the mechanism of separation can be compared to a solvent extraction, where the mobile-polar phase is one solvent and the nonpolar C-18 is the other solvent. The mechanism of separation of uncharged organics on an MPIC (polystyrene/divinylbenzene) column, however, is more closely related to adsorption than liquid-liquid solvent extraction.

In spite of the different mechanisms involved, the MPIC column can be used in much the same way that any reversed-phase HPLC can be used, except that the polystyrene/divinylbenzene in the MPIC column is stable to extremes of pH. To date, it has not been established whether some separations can be performed better on a polystyrene/divinylbenzene column or a C-18 silica column. The main applications for MPIC in the separation of

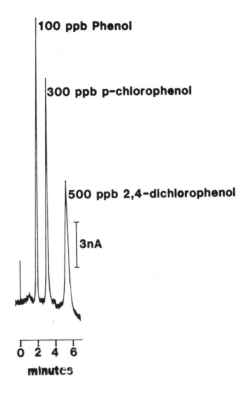

FIGURE 29.   Determination of phenols. Column: MPIC. Detection: electrochemical.

nonionic compounds is the determination of organic additives to corrosive plating solutions. Quite often these additives contain detergents similar to Triton® X-100. The general structure for this class of detergents, called alkyl phenol polyethoxylates is as follows:

$$R_1 \quad O-(CH_2-CH_2-O)_n-CH_2CH_2OH$$

$$R_1 = \text{n-nonyl or 1,1,3,3-tetramethyl butyl}$$

Other Triton® and Igepal® detergents are quite similar to the Triton® X-100 except that they have either a different number of repeating $-CH_2CH_2O-$ units and/or a straight chain $C_9H_{19}$ group on the phenyl. None of these detergents are pure. They all contain a mixture of oligomers, i.e., Triton® X-100 may contain a mix of $(CH_2CH_2O)_5$, $(CH_2CH_2O)_6$, $(CH_2CH_2O)_7$, and $(CH_2CH_2O)_8$. The MPIC column is capable of separating these oligomers. It is also capable of separating low molecular weight detergents that have the same number of $(CH_2CH_2O)$ units, but differ only in the alkyl substituent. The retention of this class of detergents decreases as the molecular weight, or percent $(CH_2CH_2O)$, increases.

   This seems to indicate that an important mechanism for retention could be the aromatic-aromatic interaction mentioned earlier. The lower molecular weight detergents have a higher phenyl content and are retained longer on the polystyrene/divinylbenzene-based MPIC column. Phenyl content is not the only factor affecting retention. Igepal® CO-210, which has the linear $C_9H_{19}$ group, is retained longer than Igepal® CA-210, which has the branched $C_8H_{17}$ alkyl group attached to the phenyl. Thus, polarity is important. The more polar, branched $C_8H_{17}$ has weaker interactions with the polystyrene/divinylbenzene than does the less polar, linear $C_9H_{19}$.

In a different class of compounds, however, the retention (capacity factor) decreases with increasing phenyl content. Alkyl phenones are well-characterized standards for HPLC.[38] On C-18 silica columns, the capacity factor increases in the homologous series: ethano-, propano-, butano-, pentano-, and hexanophenone. This is also seen on the MPIC column. The polystyrene/divinylbenzene binds the alkyl group of the hexanophenone. Although the aromatic-aromatic interactions may be involved, it is quite likely that the primary factor is simply the polarity of the analyte. Thus, the higher molecular weight detergents are more polar than their lower molecular weight counterparts, because they have more of the polar $-CH_2CH_2O-$repeating units. The higher molecular weight alkyl phenones are less polar than their lower molecular weight counterparts, and are retained longer on the polystyrene/divinylbenzene.

Applications involving MPIC of nonionic compounds are just now beginning to appear. It has been demonstrated that methanol and acetonitrile can be used with water to make eluents for MPIC. The three dihydroxybenzenes and some other compounds were separated using a 1:1 methanol/water eluent.[38] Nonionic detergents in highly alkaline plating solutions and in acidic tin/lead baths have been determined.[38]

Thus, MPIC is potentially a very powerful and versatile technique. To date, it has not received much attention in the literature. However, this is likely to change in the near future.

## IV. AUTOMATION

As with any analytical technique, ion chromatography can be made more productive through automation. More samples can be analyzed in shorter periods of time, and less applied labor is required. Because so little time is required to load samples into an auto-sampler, automation encourages the analyst to perform analyses in duplicate, or even triplicate, to help ensure accuracy. Certainly automation is an important aid when the same type of sample is to be analyzed repetitively, for example in quality control of a single product or process. But automation is also important when analyzing hazardous samples, such as water in nuclear power plants. In such cases, safety requirements mandate the use of what is essentially a remote sensing device where human interaction is minimal and can be done at a distance by computer control. Not so obvious, however, is the tremendous assistance that automation can provide to research activities. The primary advantage of automation to research is that it enables the analyst to provide ion chromatographic data up to 24 hr/day without requiring the researcher to be present. Because versatility need not be sacrificed for automation, the researcher can generate continuous ion chromatographic data with only a few hours of applied labor. Provided that the ion chromatograph is plumbed properly, compatible analyses can be performed in a completely unattended mode. It is important, however, that the analyst understand the basic principles of ion chromatography so that compatible eluents are used in a single analysis, and so column switching can be accomplished quickly and easily.

To begin with, it is important to identify each of the individual steps performed in a manual analysis, and then describe how each step can be automated. The first step in any analysis is sampling. If the sample is already quite dilute, it may first require preconcentration. An example of this is highly purified deionized water used in steam-driven turbines in power plants. To detect parts per trillion levels of highly corrosive chloride, it is necessary to first concentrate the sample on a short separator column called a concentrator column (this short concentrator column is sometimes called a guard column). On the other hand, some samples are so concentrated that they must be diluted first. Plating baths are a prime example of this. Some need to be diluted 10,000-fold prior to analysis. Still other samples need only be filtered before injection. Surprisingly, samples as complex as coffee, milk, soil-extracts, and even urine, need only be diluted, filtered, and injected. This first step, i.e., sampling, can be automated by use of an auxiliary pump, and several sampling lines (if several samples are to be analyzed). Polymeric tubing can be connected to one automated ion-chromatographic module.[40] This module can be either a dilution module (for plating bath analyses) or a concentrator module (for analyzing water in power plants). Thus, the first step in an ion-chromatographic analysis, sampling, can be automated.

The next step in the analysis is to pre-equilibrate the desired separator column and suppressor with the proper eluent. To accomplish this there must be a way to do the following:

1.  Select the desired eluent
2.  Set the desired pump speed so that the eluent flow rate is controlled
3.  Select the proper separator column
4.  Select the proper suppressor column
5.  Control the flow of regenerant through the suppressor column
6.  Control the conductivity cell

In the pump module, there are six inlets to which six different eluent reservoirs can be connected. The selection of a particular eluent inlet is controlled by solenoid-activated switching. This is, in turn, controlled by a microprocessor or a computer. The computer can be programed to activate the appropriate solenoid so that the desired eluent (or eluents in gradient elution) is/are selected.

The eluent flow rate is controlled by the speed of the analytical pump. Data on back-pressure and eluent flow rate is fed back to a microprocessor. The microprocessor uses the data to modify the pump speed until a constant flow rate is achieved. In practice, a small light is turned on, indicating to the analyst that the pump is ready. Similarly, if a gradient elution is desired, the microprocessor, or computer, can automatically select the proper combination of eluents along with the desired flow rate. At the end of the gradient program, the eluent can be returned to its original concentration and the separator column re-equilibrated so that the next sample can be properly analyzed.

The computer is also used to select the proper separator column. There is a valve, called the A valve, to which two different separator columns can be connected. It is important, however, that the two different separators be used with compatible eluents. For example, large, hydrophobic anions can be separated using the AS-5 and an isocratic bicarbonate/carbonate eluent. Aliphatic carboxylic acids can be separated on an ICE column using an acidic eluent such as 1 m$M$ HCl. If an ICE AS-1 and an AS-5 are both connected to the same A valve, it would be necessary to switch eluent from carbonate (for the AS-5) to HCl (for the ICE). In the process of switching eluents, some of the carbonate may be acidified by the HCl, and carbon dioxide gas-bubbles may be produced. These gas bubbles could disrupt regular eluent flow if they reach the analytical pump. To avoid this problem, it would be better not to connect the AS-5 and ICE to the same A valve. Instead it would be better to connect an ICE and a cation-separator column to the same A valve.

In fact, it was reported that a CS-1 and an HPICE AS-1 column can be connected to a single valve.[6] The CS-1 column was used to separate monovalent cations using a 5 m$M$ HCl eluent. The HPICE AS-1 was used to separate weak acids using a 1 m$M$ HCl eluent. On the second system of the dual-system ion chromatograph, an AS-4 and an AS-5 column were connected to this single valve. The same bicarbonate/carbonate eluent was used with both columns. The AS-4 was used to separate the more weakly retained, smaller, hydrophilic anions such as chloride, nitrate, and sulfate. The AS-5 was used to separate anions from the more strongly retained chromate.[6] In both cases, the desired column was selected by switching the A valve off or on. On one system, the AS-4 column was selected when the

**Table 3**
**AI300 DATA SYSTEM®**

**Control parameters[a]**

| Event (#) | Time | CIM Event descpt. | Event code |
|---|---|---|---|
| 1 | 0.0 | E1 ON | E1 |
| 2 | 0.0 | FLOW = 2.0 | F2.0 |
| 3 | 0.0 | LOAD | L |
| 4 | 0.0 | CELL ON | CC |
| 5 | 0.0 | OFFSET OFF | -CO |
| 6 | 0.0 | VALVE B ON | B |
| 7 | 0.1 | RLY1 ON | RL1 (sampler on) |
| 8 | 0.5 | RLY1 OFF | -RL1 (sampler off) |
| 9 | 1.0 | RANGE = 10.0 | CR10 |
| 10 | 1.0 | OFFSET ON | CO |
| 11 | 1.5 | INJECT | I |
| 12 | 1.5 | MARK | CM |

[a]Typical timed events file.

A valve was on and the AS-5 was selected when the A valve was off. On the other system, the CS-1 column was selected when the A valve was on and the HPICE AS-1 was selected when the A valve was off.

It is equally important, however, to select the proper suppressor. The CS-1 column uses a different suppressor than the HPICE AS-1. In a previous report,[6] the packed-bed ICE suppressor was connected directly to the outlet of the HPICE AS-1. The effluent coming out of the ICE suppressor was directed to a B valve. In this report, a cation fiber suppressor was also connected to the B valve, but it was connected through a separate port. In this way, when the A valve was on, the CS-1 and the cation fiber suppressor were both selected. When the A valve was off, the HPICE AS-1 and the ICE suppressor were both selected. Regardless of whether the A valve was off or on, the effluent coming out of the B valve was directed to a single conductivity detector.

The AS-4 and the AS-5 can both use the same suppressor column. Thus, only one anion micromembrane suppressor, or one anion fiber suppressor, need be connected to the B valve. In either system, the B valve serves the same function, i.e., it turns off and on the flow of regenerant. When the B valve is on, the regenerant flow is turned on. When the B valve is off, the regenerant flow is blocked.

The final parameter that is controlled by the ion chromatograph's computer or microprocessor is the conductivity cell. The computer can automatically set the conductivity cell to the desired scale, and can even change the scale in the middle of a chromatogram. This could be useful if determining a trace anion in the presence of a major anion. The conductivity detector can be set at a sensitive scale, e.g., 1 $\mu$S, when the trace anion elutes off the separator column. The detector can then be set to a less sensitive scale, e.g., 30 $\mu$S, when the major anion elutes. The computer can also automatically reset the conductivity detector to zero by turning the offset on. This is usually done just prior to sample (or standard) injection. Upon sample injection, the computer begins recording the chromatograph. The analyst can view the chromatograph in real time, or store it on a hard disk to be recalled later.

An example of a set of computer commands that can be used to automatically control a Series 2000® ion chromatographic analysis is shown in Table 3. The computer in this case is called an Auto Ion (AI) 300 Data System®. The computer commands are broken down into individual events. In the first event, the desired eluent (El) is selected. The second event sets the flow at 2.0 mℓ/min, and turns on the analytical pump. Thus, the first two events accomplish the task of equilibrating the separator column with the desired eluent at the appropriate flow rate. The third event sets the load/inject valve to the load position. In this way, the sample can be loaded into the injection loop. Next, it is necessary to prepare the conductivity cell. Event 4 turns the cell on. Event 5 turns the offset off. Now, the conductivity cell has been preset. Next, it is necessary to turn the regenerant flow on. Event 6 turns valve B on, starting the flow of regenerant. It should be noted that even though these six events have been described in sequential order, they all occur simultaneously at time 0.0.

The next step is to activate the autosampler so that the sample can be pumped into the injection loop. Thus, the seventh event, at time 0.1, turns relay 1 on. Relay 1 controls the autosampler. When it is turned on, an auxiliary pump turns on and pumps sample into the injection loop. The sample is pumped into the injection loop for 0.4 min. At 0.5 min, relay 1 is turned off, shutting off the auxiliary-sample pump. The sample is now loaded and is ready to be injected. At 1.0 min, event 9 sets the scale, or range, of the conductivity detector to 10 μS full scale. Also at 1.0 min, event 10 turns the offset to the conductivity detector on. The conductivity detector will now indicate 0.0 μS. That is, the background conductivity has been automatically subtracted. At 1.5 min, event 11 switches the load/inject valve to the inject position. The sample is now injected. Also at 1.5 min, a signal, or mark, is sent to indicate to the computer that it is to start recording chromatographic data. If, for example, the seven common anions are being separated in 8 min, the computer can be programed to stop collecting chromatographic data after 8 min. At this time, a counter is advanced so that the computer can keep up with the number of standards or samples that have been injected.

The computer can be programed so that the name of each ion can be correlated with its retention time. The program will instruct the computer to assign a portion of its memory to a variable called F (for fluoride). The program will indicate that an increase in voltage is due to the conductivity of fluoride at 1.2 min. The program will instruct the computer to either measure the peak height or area of the peak at 1.2 min, and store it in the area of memory associated with the variable F. Similarly, areas in computer memory can be assigned to the variables $C_\ell$, $NO_2$, Br, $NO_3$, $HPO_4$, and $SO_4$. The program will instruct the computer to measure peak heights (or areas) of peaks at 1.7, 2.1, 3.8, 4.2, 5.6, and 7.2 min, for example. The program will also instruct the computer to assign the concentrations of these seven anions in the first standard injected. For example, if the first standard had 0.5 ppm of each anion, the regions in the memory corresponding to the variables F, $C_\ell$, $NO_2$, Br, $NO_3$, and $SO_4$ will have not only the measured peak heights, but also the concentration 0.5 ppm. If four more standards containing, for example, 1.0, 1.5, 2.0, and 2.5 ppm are to be injected, the computer can be programed in an appropriate manner. The regions of the memory corresponding to standards 2, 3, 4, and 5 will then contain the peak heights (or areas) of the 1.0, 1.5, 2.0, and 2.5 ppm standards. After the fifth standard has been completed, the computer program can be written so that the data be analyzed by a linear least squares fit. The data consisting of the peak heights for each concentration of each anion will be used to obtain a slope and intercept four straight lines. In other words, the computer will provide seven linear plots of peak height vs. concentration. It will store the slopes and intercepts of each line for each anion.

The program can be written so that the computer assigns the next sections in memory to the names of the samples to be analyzed. For example, if tap water, well water, deionized water, rain sample, and snow sample are all to be analyzed, the computer will use five areas of memory to store the names of the five samples. If some of these samples have been diluted, the dilution factors can also be stored in memory. The computer would use the

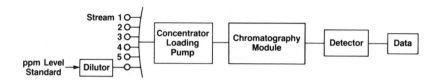

FIGURE 30.    Block diagram of Series 8000® for trace-(ppb) level analysis.

calibration data from the five standards, together with the dilution factors, to automatically calculate the concentrations of all the anions in the water samples.

It is also quite easy to perform completely different analyses. For example, one can determine chloride and sulfate in chromic acid using an AS-5 column and the same bicarbonate eluent that was used with the water samples (which used the AS-4A column). In the previously discussed automated analyses of water samples, the A valve was on, selecting the AS-4A separator column. After the standards and samples had been analyzed, the computer could be programed so that it turns the A valve off, thus selecting the AS-5 column. The program could then include 15 min of pumping eluent through the AS-5 so that it will be adequately equilibrated before injecting the first standard. The computer program would also include instructions that the 11th injection is the first standard for the new analysis with the AS-5 (the first ten injections were five standards and the five water samples). The program would instruct the computer to reserve portions of its memory for the variables $C_\ell$, $SO_4$, and $CrO_4$. The program would instruct the computer to measure the peak heights (or areas) of the peaks at 1.5, 3.9, and 13 min for each standard. The program would also indicate the concentrations in each standard. If the analyst wants to inject five standards, the computer could be programed to perform a least squares fit after the last standard was injected. The computer would then use this calibration data, together with all dilution factors, to calculate the concentrations of chloride, sulfate, and chromate in each chromic acid sample.

The way these analyses were described, all of the steps were automated except the sampling steps. The analyst would still be required to collect the sample and, if necessary, dilute it. Using the Series 8000® ion chromatograph, however, even the sampling step can be automated. A block diagram of the Series 8000® is shown in Figure 30. This diagram illustrates the most common application for the Series 8000®, analysis of ultra-pure water for trace (ppb)-level ions. This particular Series 8000® can collect five different samples (streams), and contains a separate line for ppm-level standard. The standard is diluted in a dilution module. The diluted standard and the sample streams are then loaded onto a concentrator column. The concentrator column is simply a short separator column. When present in deionized water instead of bicarbonate/carbonate eluent, anions such as chloride, nitrate and sulfate will be strongly retained on the concentrator column. These anions can then be eluted by switching to the standard bicarbonate/carbonate. The chromatography module, detector, data acquisition and reduction are all controlled as described previously, in the example using the Series 2000® ion chromatograph.

An example of how the chromatography modules in a four-channel Series 8000® ion chromatograph can be used, is shown in Figure 31. In one system, a computer-interface module (CIM) is used to control anion and monovalent cation analysis. It is quite easy to program a computer as described in the example given previously for a Series 2000® ion chromatograph. The proper eluents, columns, and detector sensitivity can be selected as described previously for the Series 2000®. A second computer-interface module can be used to control analysis of ultra-pure water for divalent cations and organic acids. Thus, in this example, ultra-pure water can be analyzed for inorganic anions, mono- and divalent cations, and aliphatic carboxylic acids, using one automated instrument. By automating the sampling step, accuracy and, in some cases, safety, is improved. Accuracy is improved because it is

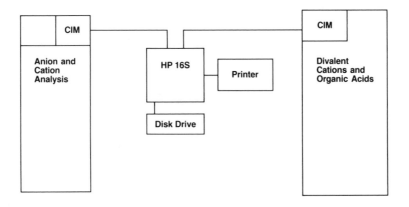

FIGURE 31.  Series 8000® four channels.

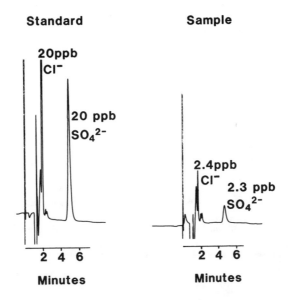

FIGURE 32.  On-line analysis of DI water.

quite difficult to sample ultra-pure water manually without introducing several ppb of chloride contamination. When analyzing water in nuclear power plants, safety is improved because radioactivity levels often mandate remote sampling to prevent exposing the analyst to harmful levels of radioactivity.

A sample ion chromatograph showing on-line analysis of ultra-pure deionized water is shown in Figure 32. The precision of the automated analysis of water for ppb-levels of anions is summarized in Table 4. The relative standard deviations obtained (1.0 to 2.2%) indicate good precision. Through the use of the concentrator column, it is apparent from these chromatograms that ppb-levels of anions can be easily detected.

Although the Series 8000® ion chromatograph is used frequently for trace ion analysis, it has also been used to analyze plating baths which contain levels of ions. Instead of a concentrator module, the Series 8000® has a dilution module. Just as in the previously described examples, the Series 8000® can be equipped with the columns and eluents needed to perform the desired analyses.

To fully utilize the versatile capabilities of the Series 8000®, it is important for the analyst

**Table 4**
**SERIES 8000®**
**PRECISION TRACE**
**LEVELS[a]**

|  | Avg. ppb | %RSD |
|---|---|---|
| $Cl^-$ | 2.97 | 1.9 |
| $NO_2^-$ | 9.9 | 2.2 |
| $Br^-$ | 9.75 | 1.4 |
| $NO_3^-$ | 10.25 | 1.8 |
| $HPO_4^{2-}$ | 15.0 | 1.0 |
| $SO_4^{2-}$ | 15.1 | 1.2 |

[a]40 Analyses.

to understand the basic principles of ion chromatography. It is also important to know the capabilities of the various separator columns so that the Series 8000® can be properly equipped.

## REFERENCES

1. **Franklin, G. O.,** Development and application of ion chromatography, *Am. Lab.,* 6, 65, 1985.
2. **Yan, D.-R. and Schwedt, G.,** Spurenanalytik von Aluminium und Eisenmittels Ionenchromatographie und post-chromatographischer Umsetzung. (Trace analysis of aluminium and iron by ion chromatography and post column derivitization), *Fresn. Z. Analyt. Chem.,* 320, 252, 1985.
3. **Heberling, S. S.,** Recent advances in metals determination by ion chromatography, Pittsburgh Conference, 1986.
4. **Cheng, K. L., Ueno, K., and Imamura, T.,** *CRC Handbook of Organic Analytical Reagents,* CRC Press, Boca Raton, Fla., 1982, 195.
5. **Langer, H. G. and Bogucki, R. F.,** The chemistry of tin(II) chelates, *J. Inorgan. Nucl. Chem.,* 29, 495, 1967.
6. **Smith, R. E. and Smith, C. H.,** Automated ion chromatography. Applications to the analysis of plating solutions, *Liq. Chromatogr. Mag.,* 3, 578, 1985.
7. **Pohl, C., Haak, K., and Fitchett, A.,** Determination of metal EDTA complexes via ion chromatography, Pittsburgh Conference, 1984.
8. **Diehl, H.,** *Quantitative Analysis,* Oakland Street Science Press, Ames, Iowa, 1970.
9. **Rocklin, R. D.,** Determination of gold, palladium, and platinum at the parts per billion level by ion chromatography, *Anal. Chem.,* 56, 1959, 1984.
10. **Haak, K. K. and Franklin, G. O.,** Rapid analysis of gold, copper, and nickel plating bath chemistry by ion chromatography, *Am. Electr. Soc. Conf. Proc.,* I-6, 1983.
11. **Fitchett, A. and Woodruff, A.,** Determination of polyvalent anions by ion chromatography, *Liq. Chromatogr. Mag.,* 1, 48, 1983.
12. **Pacholec, F., Rossi, D. T., Ray, L. D., and Vazopolos, S.,** Characterization of a phosphonate based sequestering agent by ion chromatography, *Liq. Chromatogr. Mag.,* 3, 1068, 1985.
13. **Roth, M.,** Chemical fluorescence reaction for amino acids, *Anal. Chem.,* 43, 880, 1971.
14. **Thomas, A. J.,** Developments in the fluorimetric detection of amino acids, in *Amino Acid Analysis. Methods of Enzymology,* Rattenburg, J. M., Ed., John Wiley & Sons, New York, 1981.
15. **Hirs, C. H. W.,** Amino acid analysis, *Methods Enzymol.,* 47, 3, 1977.
16. **Pihlar, B. and Kosta, L.,** Amperometric determination of cyanide by use of a flow through electrode, *Anal. Chim. Acta,* 114, 275, 1980.
17. **Pihlar, B., Kosta, L., and Hristovski, B.,** Determinations of cyanides by continuous distillation and flow through analysis with a cylindical amperometric electrode, *Talanta,* 26, 805, 1979.
18. **Rocklin, R. D. and Johnson, E. L.,** Determination of cyanide, sulfide, iodide, and bromide by ion chromatography with electrochemical detection, *Anal. Chem.,* 55, 4, 1983.

19. **Pohlandt, C.,** Determination of cyanides in metallurgical process solutions, *S. Afr. J. Chem.,* 37, 133, 1984.
20. **Bond, A. M., Heritage, I. D., Wallace, G. G., and McCormick, M. J.,** Simultaneous determination of free sulfide and cyanide with electrochemical detection, *Anal. Chem.,* 54, 582, 1982.
21. **Wang, C.-Y., Bunday, S. D., and Tartar, J. G.,** Ion chromatographic determination of fluorine, chlorine, bromine, and iodine with sequential electrochemical and conductometric detection, *Anal. Chem.,* 55, 1617, 1983.
22. **Koch, W. F.,** The determination of trace levels of cyanide by ion chromatography with electrochemical detection, *J. Res. Natl. Bur. Stand.,* 88, 157, 1983.
23. **Tarter, J. G.,** The determination of non-oxidizable species using electrochemical detection in ion chromatography, *J. Liq. Chromatogr.,* 7, 1559, 1984.
24. **Knapp, D. R.,** *Handbook of Derivitization Reactions,* John Wiley & Sons, New York, 1979, 539.
25. **Rocklin, R. D. and Pohl, C. A.,** Determination of carbohydrates by anion exchange chromatography with pulsed amperometric detection, *J. Liq. Chromatogr.,* 6, 1577, 1983.
26. **Scobell, H. D., Brobst, K. M., and Steele, E. M.,** Automated liquid chromatographic system for the analysis of carbohydrate mixtures, *Cereal Chem.,* 54, 905, 1977.
27. **Fitt, L. E., Hassler, W., and Just, D. E.,** A rapid and high resolution method to determine the composition of corn syrup by liquid chromatography, *J. Chromatogr.,* 187, 381, 187.
28. **Ladish, M. R., Heubner, A. L., and Tsao, G. T.,** High speed liquid chromatography of cellodextrins and other saccharide mixtures using water as the eluent, *J. Chromatogr.,* 147, 185, 1978.
29. **Yang, R. T., Milligan, L. P., and Mathison, G. W.,** Improved sugar separation by high performance liquid chromatography using porous microparticle carbohydrate columns, *J. Chromatogr.,* 209, 316, 1981.
30. **Sinner, J. and Puls, J.,** Non-corrosive dye reagent for detection of reducing sugars in borate complex ion-exchange chromatography, *J. Chromatogr.,* 156, 197, 1978.
31. **Edwards, P. and Haak, K. K.,** A pulsed amperometric detector for ion chromatography, *Am. Lab.,* April, 1983.
32. Dionex Corporation, Methods development in anion mobile phase ion chromatography, 1982.
33. **Iskandarani, Z. and Pietrzyk, D. J.,** Ion interaction chromatography of organic anions on a poly(styrene-divinylbenzene) adsorbent in the presence of tetraalkonium salts, *Anal. Chem.,* 54, 1065, 1982.
34. Dionex Corporation, Determination of ethanolamines in refinery water, *Appl. Note,* 39, 1983.
35. **Edwards, P. and Haak, K. K.,** A pulsed amperometric detector for ion chromatography, *Am. Lab.,* April, 1983.
36. **Burley, S. K. and Petsko, G. A.,** Aromatic-aromatic interaction: a mechanism of protein stabilization, *Science,* 2229, 23, 1985.
37. **Kitka, E. J. and Stange, A. E.,** Phenones: a family of compounds broadly applicable to use as internal standards in high performance ion chromatography, *J. Chromatogr.,* 138, 41, 19.
38. **Davis, W. D., Smith, R. E., and Yourtee, D.,** Analysis of organics in plating solutions using an ion chromatograph, *Metal Finish.,* 63, Sept. 1986.

Chapter 4

# APPLICATIONS

## I. LIFE SCIENCES

When most ion chromatographers think of ion chromatography, they think of determining inorganic anions in dilute aqueous solutions that bear little resemblance to soil samples, blood serum, or food and beverages. Moreover, few biochemists or pharmaceutical scientists have done much ion chromatography. This is unfortunate, because biochemists and pharmaceutical scientists are constantly trying to measure the concentrations of various ionic substances. For example, one of the chromatography papers from the 1986 Pittsburgh Conference that was specially selected for publication, described the separation of several sugar phosphates by ion exchange.[1] They were detected by adding a lanthanide metal, europium, which forms a UV-absorbing complex.[2] This permitted the use of a UV detector, but detection limits were on the order of 1 mg/m$\ell$. In contrast, it has also been shown that sugar phosphates can be separated by ion chromatography, and can be detected at the milligram per liter level using chemically suppressed conductivity detection.[3,4] This is but one example. There are certainly many more classes of ionic compounds that have already been separated on more conventional ion exchange resins that could be separated faster and easier with ion chromatography separator columns. The detection mode used would depend on the chemical properties of the ion. Zwitterions might be detected by their UV absorbance, their electrochemical reactivity, or by post-column reaction to form a fluorophor or a chromophore. Again, a knowledge of the various types of columns and detectors is necessary in order to develop new analytical methods rapidly and efficiently.

This chapter will describe some of the applications that have already been developed. Samples as varied as water washings of soil samples to filtered physiological fluids have been analyzed. The analytes have included common anions and cations, amino acids, carboxylic acids, and carbohydrates. In many cases, the only sample preparation required is to dilute, filter, and inject. Of course, guard columns should be used. The data from these analyses has been used to address some very real scientific and industrial problems. Large-scale production of monoclonal antibodies can be monitored with ion chromatography. The flavor of foods and beverages can be controlled by monitoring the concentrations of carbohydrates. One of the more interesting challenges that was made in the early ion chromatography literature was to use ion chromatography to study environmental genotoxicants. Compounds as varied as nitrite, sulfur dioxide, chromate, azide, hydrazine, aromatic amines, and aza arenes were identified as important potential analytes back in 1978.[5] In this paper, Sawicki quite astutely predicted that ion chromatography should be capable of determining all these compounds, even though techniques such as mobile-phase ion chromatography (MPIC) and pulsed amperometric detection would not appear for 3 to 5 years. In the near future, it is quite possible that many such compounds, as well as many compounds of interest to biochemists, will be determined routinely by ion chromatography. It is also quite possible that ion chromatographers and physical organic chemists will begin a semiquantitative structure-activity study relating the structures of genotoxicants and chemical mutagens to their chromatographic properties. Certainly such studies on other organic compounds have been performed using gas chromatography and conventional liquid chromatography. It will be useful to perform similar studies using both ion-exchange columns with their immobilized, permanent charges that may bear some resemblance to the immobilized, permanent charges on DNA. It may also be possible to correlate the ability of toxic chemicals to absorb into the soil and leach into the groundwater to their chromatographic properties. For these and

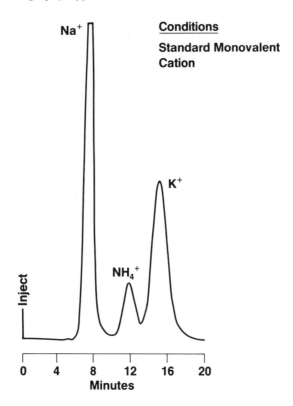

FIGURE 1.    Alkali metals in urine. Column: CS-1, Eluent: 5 m$M$ HCl.

many other reasons, it is quite possible that the life sciences represent the major growth area for ion chromatography in the near future. The number of significant analytical problems that can be solved conveniently by ion chromatography is just now beginning to be explored. In this section, some of these beginnings are described.

One of the earliest applications was in the area of urine analysis, which presents an unusually complex sample matrix. It contains ill-defined mixtures referred to as "uro-chromes". It also contains polyanions which are known to cause problems such as irreversible binding on ion-exchange resins. These include polynucleic acids and polypeptides. In spite of this, urine was analyzed for sodium, ammonium, and potassium with the original cation separator, the CS-1, and the original packed-bed cation suppressor as shown in Figure 1. For modern routine analysis, it is advisable to use a guard column and to check it periodically, in the manner described in Chapter 1 (Section XI) and illustrated in Figure 15 (Chapter 1).

In addition, it is now possible to analyze urine and many other samples for these mon-ovalent cations and for the alkaline earth metals (with calcium and magnesium being of special interest). In an earlier paper, it was shown that serum can be analyzed for calcium and magnesium using the CS-1 column.[6] Serum specimens were ultrafiltered to help protect the separator column. The samples were also acidified with HCl to remove calcium and magnesium that is bound to proteins in vivo. Ion chromatographic results were compared to atomic absorption spectroscopy results by these workers,[6] and were found to agree quite well. Concentrations were on the order of 20 mg/$\ell$ for magnesium and 100 mg/$\ell$ for calcium. This work, however, was done before the advent of the CS-3 column and gradient elution. It is now possible to separate and detect mono- and divalent cations in one run. The separator column for this analysis is the CS-3, and the suppressor column is the cation micromembrane suppressor (see Chapter 2, Section IV and Figure 23, Chapter 2 for more details).

One other important application using the CS-1 (or CS-3) is the analysis of lithium, which

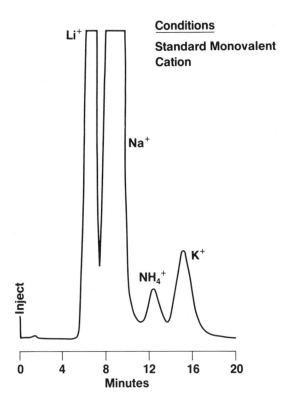

FIGURE 2.    Lithium levels in patient under treatment for depression. Column: CS-1, Eluent: 5 m*M* HCl.

is used to help treat patients suffering from depression. Lithium has no metabolites and exhibits simple first-order elimination kinetics with a half-life of about 14 to 33 hr. Lithium is usually determined in blood serum by flame photometry.[7] At steady-state, there is a fluctuation in lithium in which serum concentrations rise 15 to 30% above early morning baseline levels. The therapeutic concentration range is about 0.3 to 1.3 mmol/$\ell$. Severe toxicity occurs at concentrations above 2 mmol/$\ell$.[7] In the absence of renal damage, the lithium kinetics are predictable, and changes in serum concentrations of lithium can be used to help predict doses that should be given. Although flame photometry is capable of monitoring lithium, so is ion chromatography, as shown in Figure 2. In this particular chromatogram, ammonium and potassium can be determined. For routine lithium analysis, it is better to dilute the sample more, so that the lithium peak is on-scale. In such cases, the ammonium and potassium may be too small for accurate quantitation.

Anions were also determined on the first anion separator, the AS-1. As shown in Figure 3, the AS-1 was used to monitor phosphate and sulfate in chronic renal failure patients. Similarly, anions in an intravenous solution were determined as shown in Figure 4. One group reported a detailed analysis for sulfate in human plasma, cerebrospinal fluid (CSF), and liver.[8] As in so many other applications, these separations can now be performed faster using more modern anion separators such as the AS-4A and AS-5. Certainly, it is also much more convenient to use an anion micromembrane suppressor than the packed-bed suppressor that was used in Figures 2 and 3. Both of these separations used the most popular 3 m*M* sodium bicarbonate/2.4 m*M* sodium carbonate eluent. In some cases, other eluents are needed. For example, it is important to monitor acetate, pyruvate, and formate in Ringer's lactate solution. These three carboxylic acids are weakly retained on anion-separator columns, and would co-elute near the void volume if the bicarbonate/carbonate eluent was used.

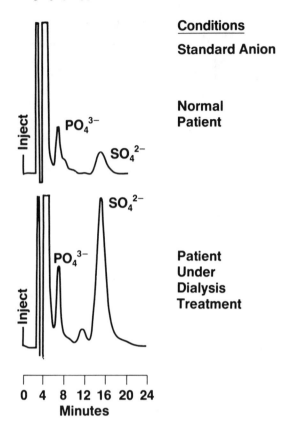

**Conditions**

**Standard Anion**

**Normal Patient**

**Patient Under Dialysis Treatment**

FIGURE 3.   Monitoring phosphate and sulfate in chronic renal failure patients. Column: AS-1.

Instead, they can be separated using a weaker eluent, 5 m*M* sodium borate, as shown in Figure 5.

One rather popular application for ion chromatography is the determination of urinary oxalate, with at least three papers and one application note having appeared.[9-12] Oxalate can cause painful kidney stones if its concentration builds up too high, but it has also been suggested that it has important properties as an anticarcinogen.[13] As suggested once by the inventor of the now famous Ames test for mutagenicity, perhaps the healthiest diet is one that is just rich enough to keep one on the verge of gout. At any rate, it is important to monitor the concentration of oxalate in patients with gout, and in any epidemeological data that might be collected in determining the degree of anticarcinogenicity of oxalic acid. Oxalate is a divalent anion, and is therefore retained rather well on anion-separator columns. In earlier work, oxalate was shown to elute after sulfate on an AS-3, or fast separator column, as it is referred to in an application note.[12] As shown in Figure 6, oxalate is well separated from the sulfate peak, which goes off the scale at about 9 min. The other major peaks that go off the scale are due to monovalent organic acids, chloride, nitrate, and phosphate. In a more recent study,[10] it was shown that ascorbic acid (Vitamin C) can be oxidized in the sodium bicarbonate/sodium carbonate, pH 10.7 eluent used in Figure 6. In most patients, this is no problem, because urinary concentrations of ascorbic acid are much lower than oxalate. However, if a patient has been taking megadoses of Vitamin C, appreciable amounts can be oxidized to oxalate. To avoid this, 0.75 m$\ell$ of sample is mixed with 0.5 m$\ell$ of 1 *M* HCl, and then diluted to 50 m$\ell$ with 0.3 *M* boric acid. Ascorbic acid is not oxidized at this low pH. The boric acid neutralizes the pH 10.7 eluent. With this bicarbonate/carbonate

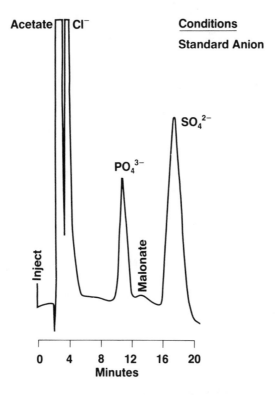

FIGURE 4.    Anions in intravenous solutions. Column: AS-1.

| Conditions | Concentrations (ppm) | |
|---|---|---|
| Standard Anion | Acetate | 20 |
| Except | Formate | 10 |
| Eluent: | Pyruvate | 45 |
| 0.005 Na$_2$B$_4$O$_7$ · 10 H$_2$O | | |

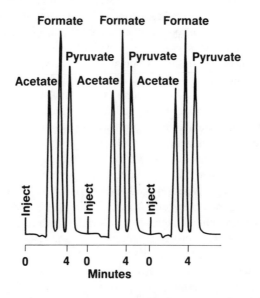

FIGURE 5.    Repeat analysis of ringer's lactate solution. Column: AS-1.

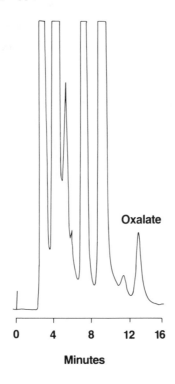

FIGURE 6.    Urinary oxalate. Column: AS-3

eluent, ascorbate elutes at the void volume and can be detected with a UV detector set at 254 nm.[10]

For a more complete profile of organic acids present in physiological fluids, other separation techniques may be needed. In their landmark paper, Rich et al.[14] described the simultaneous determination of organic acids in plasma and urine. They used both ion chromatography exclusion (with an ICE AS-1 column) and ion-exchange chromatography (with an AS-1 separator column). The two techniques were combined in a two-dimensional technique called ion chromatography/ion chromatography exclusion, or IC/ICE. In this method, the sample is first injected on the ion-chromatography exchange column (ICE AS-1) and, at an appropriate time, a valve is switched so that the column effluent is switched to an anion-exchange, AS-1, column. In this particular paper, lactate and succinate needed to be separated, but they co-eluted at about 17 min on the ICE column. The effluent, after about 17 min, was switched to an anion-separator column, AS-1, which easily separated the lactate and succinate. In addition, chloride, nitrate, and sulfate were co-eluted on the ICE column, but were separated after subsequent chromatography on a coupled anion separator, the AS-1. Coupled chromatography was also used by Rich et al.[14] to determine the concentration of an aromatic acid, vanillylmandelic acid (VMA). In this case, the sample was first injected on the anion separator, and then the fraction containing VMA was rechromatographed on an ICE column. This technique is abbreviated IC/ICE. The VMA, together with some other aromatic compounds, co-eluted between 12 and 17 min on the IC separator. This fraction was reinjected on an ICE column, and the VMA was well separated from the other aromatics. The accuracy of the method was verified by comparing results from the ion chromatography to results from the standard method of Pisano et al.[15] The ion-methods chromatographic method gave slightly lower values for urinary VMA (1.4, 4.9, and 4.0 mg/$\ell$ compared to 2.4, 5.2, and 6.0 mg/$\ell$ by the Pisano method). This was attributed to the lack of specificity of the Pisano method.

One of the major contributions of the paper by Rich et al.[14] was the description of sample preparation. Instead of the time-consuming sample pretreatments needed for gas and conventional liquid chromatography, all that was needed was centrifugation, filtration, and dilution. The plasma samples required centrifugation at 2600 rpm for 5 min. The plasma was decanted, and diluted with an equal portion of acetonitrile. After a brief vortex mixing, the clear supernatant could be analyzed. Urine was analyzed by simply adding 10 m$\ell$ of 6 $M$ HCl, and storing at 4°C. An aliquot was diluted tenfold before injection.

One of the interesting things about this paper is that it describes a series of difficult analyses using some of the oldest columns available. It would be interesting indeed, to perform multicomponent ion analysis of physiological fluids using the collection of new separator columns and gradient elution that is now available. There are at least three anion separator columns (AS-4A, AS-5, and AS-6) with distinctly different properties. There are at least two different ICE columns (HPICE AS-1 and AS-5) that could also be used to characterize the anions present. These columns have already been shown to be capable of separating six monocarboxylic acids (Figure 12, Chapter 2), six dicarboxylic acids (Figure 13, Chapter 2), nucleoside monophosphates (Figure 6, Chapter 2), and nucleoside triphosphates (Figure 4, Chapter 2). It is very likely that these new separator columns, together with new detection methods, will be used to analyze a variety of physiological fluids and biochemical samples in the near future.

Physiological fluids are not the only sample that require analysis for organic acids. When bacteria such as *E. coli* are in a fermentation medium, acetic acid will be produced. To help monitor the fermentation, ICE can be used to periodically analyze the broth as shown in Figure 7. In a similar application, microbiologists used ICE to determine metabolic patterns of the sulfate-reducing bacteria, Desulfovibrio.[16] These workers monitored the concentrations of sulfate, lactate, acetate, and carbonate. Certainly, microbiologists have been involved in applications that are both scientifically and commercially important. One of the most important products, vinegar, is produced through a fermentation process. To help characterize this process, organic acids in corn vinegar were compared to those in cider vinegar as shown in Figures 8 and 9. Lactic and succinic acids are more concentrated in this sample of cider vinegar, whereas tartaric acid is more concentrated in corn vinegar. Also of interest are the carbon sources that are used in producing these acid metabolites. Sugars and alcohols can be separated on an ICE column and be detected by pulsed amperometric detector (see Chapter 3, Section III and Figures 13 to 15, Chapter 3). Corn and cider vinegars are compared in Figures 10 and 11. Again, significant differences are seen. Corn vinegar is much sweeter due to its elevated fructose concentration, whereas cider vinegar can have appreciable levels of glycerol and ethanol.

The concentrations of sugars in food and beverage products are very important. The market for sweetened soft drinks is so large that it is very important that the proper proportions of the very important sugars be accurately monitored. A sample analysis, the determination of glucose, fructose, and sucrose, is shown in Figure 12. In this case, however, ion chromatography exclusion (ICE) was not used for the separation. Instead, the sugars were separated as anions. These and other sugars do have an ionizable −OH with pKa values ranging from 12 to 14. Thus, they are anions in 0.16 $M$ KOH and can be separated on the relatively high-capacity AS-6 column that was used in Figure 12. Even samples as complex as potato chips require accurate-quality flavor control. Because the AS-6 column is so rugged, it can be used to separate glucose, fructose, lactose, and sucrose from other sugars that are present as shown in Figure 13.

Another class of compounds that is easily determined by ion chromatography is amino acids. They can be separated as cations as low pH, or as anions at a higher pH. Classically, this often required the use of two columns, a long column for acidic and neutral amino acids, and a short column for the basic amino acids. In the ion-chromatographic methods,

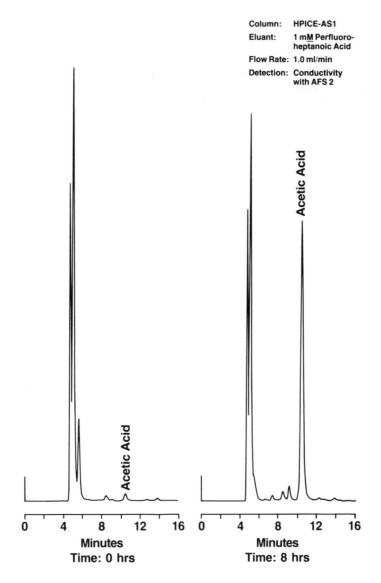

Column:     HPICE-AS1
Eluant:     1 mM Perfluoro-
            heptanoic Acid
Flow Rate:  1.0 ml/min
Detection:  Conductivity
            with AFS 2

FIGURE 7.    Fermentation broth. Column: ICE.

only one column is required. Amino acids in an intravenous solution can be separated as cations as shown in Figure 14. On the other hand, amino acids exist as anions at a higher pH, and can be separated on a novel anion separator, the AS-8, as shown in Figure 15. In both cases, detection is accomplished by post-column reaction with *o*-phthalaldehyde. The effluent off the AS-8 is directed to a post-column membrane reactor. This membrane reactor is semipermeable to the *o*-phthalaldehyde solution which bathes the outside of the membrane. The AS-8 column effluent flows countercurrent to the flow of *o*-phthalaldehyde. As the *o*-phthalaldehyde crosses the membrane, it reacts rapidly with the individual amino acids to form a fluorophor. This fluorophor is detected with a fluorescence detector.

Probably the major area in the life sciences for ion chromatography is agronomy. Soil, irrigation waters, and plants all require ion analysis. One of the earlier papers in this field described the determination of total sulfur and chloride in plant materials.[17] The plants were combusted in a Schoniger flask. In this procedure, a few milligrams of sample is weighed and then wrapped in a small piece of sample wrapper paper and placed in a flask that contains

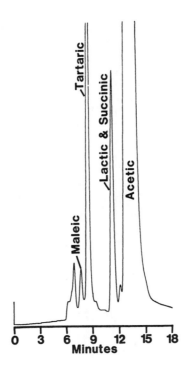

FIGURE 8.  Organic acids in corn vinegar. Column: ICE.

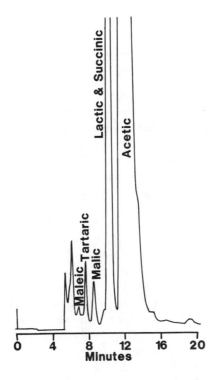

FIGURE 9.  Organic acids in cider vinegar. Column: ICE.

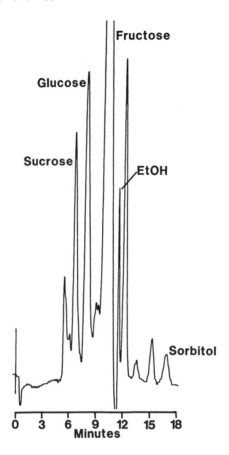

FIGURE 10.    Sugars and alcohols in corn vinegar. Pulsed amperometric detector.

about 10 mℓ of bicarbonate/carbonate eluent. The sample and paper are ignited by an infrared light and all organic chlorine and sulfur is converted to chloride and sulfate. In this report, the original separators and packed-bed suppressors were used to determine the chloride and sulfate produced. In a more recent study, total potassium, sodium, calcium, and magnesium were determined in plant materials.[18] Because the CS-1 column was used, two separate, dedicated columns were required for the analysis. Today, however, the same analysis can be performed on one column, the CS-3, in one run. In a similar application, potassium and bromate were determined in bread.[19] Potassium bromate is used as an oxidizing agent in baking bread. It is also mutagenic. The potassium was separated from other cations on the CS-1. The bromate was separated from other anions on an AS-1. If the standard sodium bicarbonate/sodium carbonate eluent had been used, bromate and chloride would have co-eluted. Instead, a weaker eluent, 3.5 m$M$ sodium tetraborate, was used. Bromate eluted in about 28 min, and chloride in 32 min. Again, however, this data is certainly valid, bromate and chloride can be separated much faster on an AS-4A column.

In another report, urea in fertilizer was determined.[20] The sample was mixed with the enzyme urease, which catalyzes the hydrolysis of urea to ammonia. After 1 hr at 20°C, HCl was added to convert the ammonia to ammonium, which can then be quantitated using a CS-1 and fiber suppressor.[20] Similarly, the CS-1 can be used to determine ethylenediamine in a veterinary livestock iodine supplement as described by Buechele and Reutter.[21] Because ethylenediamine is divalent, it is strongly retained, and requires a strong eluent to be eluted. The eluent used was 4 m$M$ HCl with 2.5 m$M$ zinc chloride. The zinc forms a complex with the ethylenediamine to effect elution. Many monovalent cations, including ammonium, and

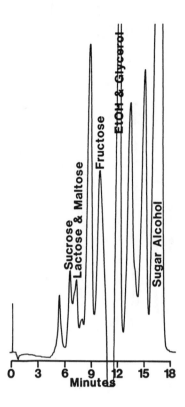

FIGURE 11.    Sugars and alcohols in cider vinegar. Pulsed amperometric detector.

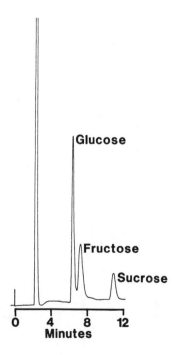

FIGURE 12.    Sugars in root beer. Pulsed amperometric detector.

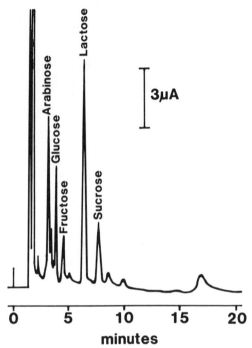

FIGURE 13.    Carbohydrates in food. Extract from flavored potato chips. Pulsed amperometric detector.

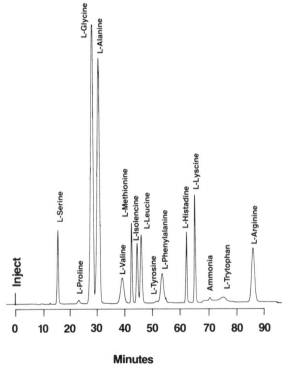

FIGURE 14.    Amino acids in i.v. solution. Ninhydrin detection. Separated as cations.

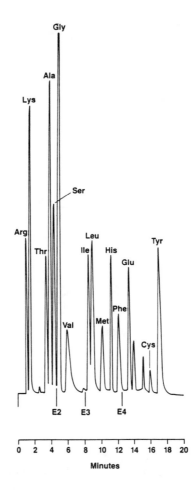

FIGURE 15.    Protein hydrolyzate by anion exchange.

alkali metals, and nine amines, were all found to co-elute in the void volume using this eluent. Another food product that has been analyzed by ion chromatography is shellfish. It is important to analyze shellfish for tetramethylammonium, a source of food poisoning. Ito et al.,[36] performed this analysis by homogenizing shellfish salivary glands in methanol. The methanol was evaporated, and the pH of the homogenate was adjusted to 5 with HCl. The fats were then extracted with ether, and the sample was diluted to 10 mℓ with water. The tetramethyl ammonium was separated from other cations on a CS-2. It was detected by chemically suppressed conductivity.

In a review of the subject, Edwards[22] described several applications of ion chromatography in analyzing food samples. Some analytes that are easily determined by ion chromatography, represented significant problems to the food chemist before the development of ion-chromatographic methods. Traditional methods for nitrate, such as the cadmium-reduction and xylenol-orange methods, are tedious, and show different results when performed by different analysts. Nitrate levels were measured in carrots and ham by both ion chromatography and by classical wet-chemical methods by the Committee of Food Industry Analysts (CFIA), a committee of the National Food Processors Association. The ion-chromatographic results were found to be quite accurate in these tests. Nitrate levels in carrots ranged from 100 to 424 ppm. Ham samples had from 91 to 121 ppm. Ion chromatography required much less time than the conventional methods. Sodium, potassium, and calcium were determined in tomato juice, canned tomatoes, canned corn, canned, puréed beef, instant oatmeal, canned

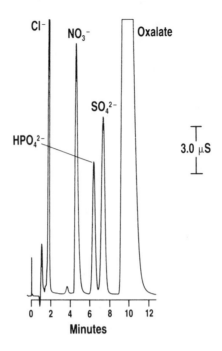

FIGURE 16.    Anions in spinach leaf. Sample preparation: homogenize, filter, dilute, and inject.

green beans, and in apple juice. Again, ion chromatography accuracy was tested by the CFIA. For the ion chromatographic method, the samples only had to be ground or homogenized, weighed, and then diluted into the 5 m$M$ HCl eluent. Sample pretreatment for the classical methods, including those from the Association of Official Analytical Chemists (AOAC), was more elaborate and time-consuming. These methods did give results that were in agreement with those obtained by ion chromatography. The transition metals, zinc and iron, were also determined in green beans. They were separated on a CS-2 column using a 10 m$M$ oxalic acid/7.5 m$M$ citric acid eluent. They were detected by a post-chemical reaction with the colorimetric reagent PAR. The same type of membrane reactor was used for the transition metals as was described above for amino acids. Unlike the amino acids, however, the transition metals were detected at 510 nm with a UV visible detector. The last method that was used in this comparison with CFIA methods was the determination of the carbohydrates, fructose and glucose. The ion-chromatographic method used was the same AS-6 column and pulsed amperometric detector used in Figures 12 to 14.

In a more recent report,[23] anions were determined in spinach leaf. A sample chromatograph of this type of analysis is shown in Figure 16. In this chromatogram, chloride, nitrate, phosphate, and sulfate can be quantitated. Oxalate is easily quantitated by simply diluting the sample further. Similarly, mono- and divalent cations can be determined simultaneously in one run using the CS-3 separator and a cation-micromembrane suppressor as shown in Figure 17. Another class of analytes that are not always thought of primarily as cations is water-soluble vitamins. As shown in Figure 18, they can be separated as cations. In this case, a strong eluent, 8 m$M$ hydroxylamine hydrochloride, pH 3, was used. The vitamins were detected at 215 nm with a UV-visible detector. In the midst of all the advances in ion-chromatography hardware, however, there is still an important place for one of the older columns, the AS-2. It is still useful in determining nitrate in samples containing high chloride levels. Although the AS-2 was originally developed for analyzing brines, it can also be used to determine nitrite levels in a high chloride-containing food, bacon, as shown in Figure

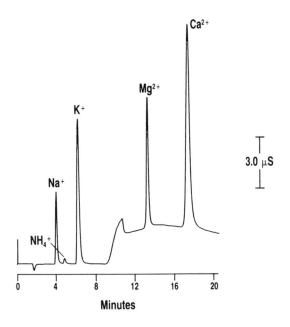

FIGURE 17. Cations in celery leaves.

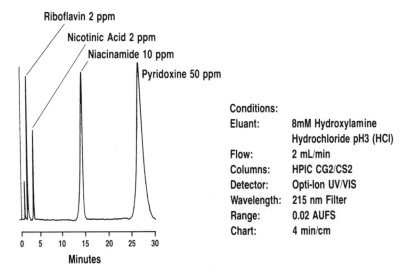

FIGURE 18. Separation of four water-soluble B vitamins.

19. Similarly, a method for determining iodide in milk has been available for some time.[24] As shown in Figure 20, iodide can be separated from other anions using an MPIC column and an eluent consisting of 2 m$M$ tetrabutyl ammonium hydroxide, 1 m$M$ sodium carbonate, and 20% acetonitrile. The iodide is detected on a silver electrode, to which $+0.3$ V are applied in an electrochemical detector. Another analysis that was developed recently is the determination of sulfite in food samples. The sample is homogenized and the sulfite is extracted by flash-distillation using an apparatus like that shown in Figure 21. To prevent oxidation of the sulfite, the apparatus is purged with nitrogen gas, and the cold trapping solution contains formaldehyde. The formaldehyde forms a complex with sulfite which stabilizes the sulfite to oxidation, and enhances its separation on an AS-4 as shown previously.[25]

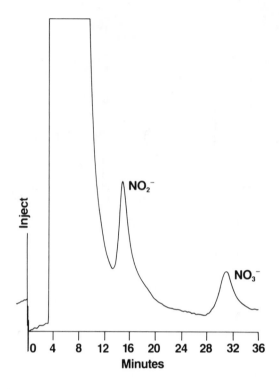

FIGURE 19.    Nitrite and anitrate in processed bacon. Conductivity detection. Column: AS-2.

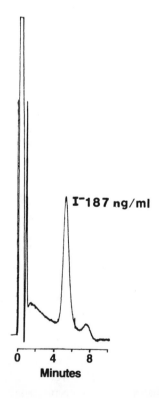

FIGURE 20.    Iodide in whole milk. Electrochemical detection.

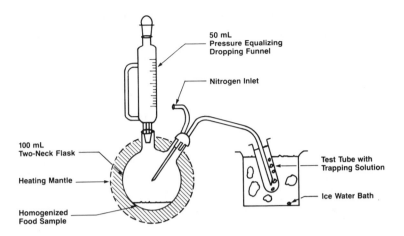

FIGURE 21.    Flash distillation apparatus for IC method for sulfite sample preparation.

## II. ENVIRONMENTAL SCIENCES

Chemically suppressed conductivity had immediate applications in the field of chemically suppressed ion chromatography. Environmental chemists are often required to analyze many samples taken throughout regularly spaced areas in the environment. This enables them to perform a rapid survey of the environment for pollutants for source assessment. However, the problem is compounded by the requirement for multicomponent determinations in each sample. The component ions can be quite diverse. The list of analytes includes common anions such as chloride, nitrate, phosphate, and sulfate, and less common organics such as ethanolamines. Because the first applications in ion chromatography were the determination of common anions, it is this area that has been most thoroughly developed. Today, ultrapure deionized water can be analyzed for parts per trillion levels of anions after being preconcentrated on a short anion guard column. In addition, soil samples can be extracted with aqueous solutions and analyzed for ionic components. The earliest application of ion chromatography to environmental analysis, however, was the analysis of air samples. Thus, ion chromatography can be used to analyze atmospheric samples, wastewaters, soil, and even highly purified water for ions that have an environmental impact.

The first two books written on ion chromatography were conference proceedings. In 1977 and 1988, symposia were held at the U.S. Environmental Protection Agency in Research Triangle Park, N.C. The proceedings[26,27] describe primarily the analysis of atmospheric samples, including aerosols and particulate matter. Environmental samples were analyzed for ammonium, chloride, nitrate, phosphate, and sulfate. Potential sources of pollution as diverse as diesel exhaust fumes and air-bags used in automobiles, require examination. The ion-chromatographic methods were verified by comparison to other analytical methods that were routinely used in 1977 and 1978. The separator and suppressor columns used then are now considered to be almost obsolete. Now, the common anions require only 7 min for baseline separation, where 12 min were required in 1977. Even in 1978, however, environmental chemists talked about analyzing environmental samples for less common genotoxic substances such as nitrite, sulfur dioxide, azide, hydrazide, hydroxylamine, aromatic amines, and aza arenes.[28] With modern technology that includes alternate separation and detection methods, many such compounds can now be determined.

One of the best-publicized environmental problems is acid rain. Highly industrialized areas can produce measurable levels of sulfuric and nitric acids that can be harmful to environments that are downwind. Oxides of sulfur and nitrogen can be carried several hundreds of miles. They are converted to their acid forms in rain and snow, lowering the pH of lakes and streams even in remote areas. Ion chromatography is quite useful for quantitating not only the major anions, fluoride, chloride, nitrate, and sulfate, but also the monovalent cations, sodium, ammonium, and potassium. These can all be determined simultaneously in one sample injection. The ion chromatographic procedure is almost the same as that first developed by Small and Stevens. It has been used by numerous workers to determine the ionic contents of natural waters (rain, ice, snow, and drinking water) from Japan to Germany, and even in Bakersfield, Calif.[29-42] Acid-rain samples from the MAPS-3 region of the northwestern U.S. was found to have 2 mg/ℓ sulfate, 3 mg/ℓ nitrate, and 0.2 mg/ℓ chloride in 1980. One research group from the Federal Republic of Germany, has analyzed drinking water and found 0.1 ppm fluoride, 30 ppm chloride, less then 0.1 ppm nitrite, less than 0.5 ppm phosphate, 6 ppm nitrate, and 35 ppm sulfate.[37] In contrast, snow and ice samples taken from Antarctica were found to have about 2 ng/ℓ ammonium, 2 ng/ℓ potassium, 15 ng/ℓ nitrate, 70 ng/ℓ sulfate, 22 ng/ℓ sodium, and 45 ng/ℓ chloride. These samples can probably be taken as "natural" levels of anions in water, because samples can be taken by drilling-down to levels of ice that were deposited long before the industrial revolution and large-scale burning of fossil fuels began. Thus, it is interesting to note that the ion content of Antarctica snow and ice is about 100 times lower than in the more industrial areas of the northern hemisphere. Of course, sulfate is an important ion to observe in these studies because it is a measure of the sulfur content in the fossil fuels being burned. It is about 500 times higher in Germany than in the Antarctic. Nitrate is another important anion because it can be reduced to nitrite which can be converted in the body to carcinogenic nitrosamines.[28]

In another study, yearly levels of atmospheric deposition of sulfate, nitrate, ammonium, calcium, and potassium in forest areas were measured.[43] These workers sampled dry deposition, precipitation before and after its interception by the forest canopy, and determined the concentrations of airborne particles and vapors. These values were used to estimate atmospheric input to the canopy and forest floor. Their data supports the hypothesis that dry deposition plays a major role in atmospheric inputs to deciduous forests. These depositions noticeably affect ion uptake and loss in the forest. They estimated that atmospheric deposition contributes 40% of the annual woody increment of calcium and 100% of the sulfur. It has been suggested that the effects of excess sulfur, nitrogen, and trace-metal atmospheric deposition are already being manifested in high-elevation forests in the eastern U.S. and Europe.[44]

Chemically suppressed ion chromatography was compared to conventional analytical methods in a round-robin study in order to obtain acceptance for drinking-water analysis under the Safe Drinking Water Act.[45] To obtain nationwide equivalency, the methods were compared at two independent laboratories. Fluoride was the only SDWA constituent analyzed which did not show overall agreement between methods. In samples of very hard water, the calcium carbonate concentration can reach 250 mg/ℓ. This produces a sizable carbonate peak which overlapped with the fluoride, producing a positive bias. Chloride concentrations ranged from 5 to 240 mg/ℓ, nitrite from 0.04 to 5.6 mg/ℓ, phosphate from 0.1 to 5.5 mg/ℓ, nitrate from 0.1 to 15 mg/ℓ, and sulfate from 4 to 390 mg/ℓ. In every case, conventional EPA or APHA methods were used for comparison with ion-chromatographic results. None of these methods, however, was capable of the simultaneous multicomponent analysis that ion chromatography can produce. To verify the accuracy of the ion-chromatographic method, drinking-water samples were analyzed by standard methods and by ion chromatography. The results obtained by ion chromatography were plotted against the results obtained by the conventional method. The results are summarized in Table 1. The correlation coefficients range from 0.993 to 0.996, indicating that the ion-chromatographic method yielded similar results to those obtained by conventional methods.

**Table 1**
## SUMMARY OF DRINKING-WATER RESULTS

|  | Range |
|---|---|
| Fluoride | |
| $F^-$ (probe) = 0.948 F (IC) + 0.057 | 0.993 |
| Chloride | |
| $Cl^-$ (potentiometric) = 1.01 Cl (IC) − 1.3 | 0.996 |
| Nitrite | |
| $NO_2^-$ (colorimetric) = 0.98 $NO_2^-$ (IC) − 0.018 | 0.986 |
| Phosphate | |
| $PO_4^{3-}$ (colorimetric = 1.05 $PO_4^{3-}$ + 0.066 | 0.977 |
| Nitrate | |
| $NO_3^-$ (colorimetric) = 0.93$NO_3^-$ (IC) + 0.17 | 0.996 |
| Sulfate | |
| $SO_4^{2-}$ (turbidimetric) = 0.98 $SO_4^{2-}$ (IC) = 0.25 | 0.993 |

These results were instrumental in ion chromatography being accepted by the Environmental Protection Agency (EPA) as the method of choice for determining chloride, phosphate, nitrate, and sulfate in precipitation samples in an EPA quality-assurance manual. In addition, the American Society for Testing and Materials (ASTM) has incorporated ion chromatography into standard test procedure for the analysis of rain.[41] The ion chromatograph used in the ASTM method uses high-technology polymers with the only metal contacted being located in a small area in the analytical pump head. The injection valve normally contains a fixed volume loop of 100 $\mu\ell$ volume. For subparts per billion detection, this loop is replaced by a concentrator column. In this way, volumes significantly greater than 100 $\mu\ell$ can be concentrated before injection on a separator column. Precisions on the order of 1% (relative standard deviation) were reported for anions. Not only anions, but also monovalent cations, can be determined in the ASTM method. These include sodium, ammonium, and potassium as well as ethyl amine and triethyl amine. Of course, these are separated on a cation-separator column. For the divalent metals, calcium and magnesium, the older procedure using a dedicated cation-separator column and $m$-phenylenediamine eluent are incorporated into the ASTM procedure. This is because the CS-3 column, which permits simultaneous mono- and divalent cation separations, was not available when the ASTM method was written.

Although most of these studies utilized a concentrator column to achieve the necessary sensitivity, a report has appeared in which the ion chromatograph was modified so that preconcentration was not necessary. The most severe limit to sensitivity, the water dip, is minimized in the approach of Pyen et al.[43] They used two pulseless pumps. One was used to pump sample and the other to pump eluent ten times stronger than that normally flowing through the ion-chromatographic system. The two pumps produce two streams of solution. These two streams are mixed at a "T" fitting prior to the injection port. In this way, the samples are put into an aqueous solution with the same ionic strength of the eluent, and this minimizes the negative peak called the water dip. The analytical procedure was automated with microprocessor control, permitting fully unattended analysis. Detection limits as low as 0.01 mg/$\ell$ were reported without the need for sample preconcentration. The fluoride and chloride peaks were usually large enough to be detected using the 30-$\mu$S scale on the conductivity detector. After chloride has been eluted, the scale is switched to 1 $\mu$S to allow detection of the smaller peaks due to phosphate, bromide, nitrate, and sulfate. This is all controlled automatically by the microprocessor.

Thus, ion chromatography can be very useful in analyzing water samples. Ion chromatography has been used to analyze more than just acid-rain samples. Other major applications in analyzing water samples of environmental interest include drinking water, industrial-waste streams, bio-pond waters, ocean waters, and extracts from geological samples. One study[46] reported the determination of the seven common anions in nine different industrial waters: boiler feed, boiler, cooling, zeolite softened, softened, nitrite-treated, city make-up, well water, and waste effluent. These analyses were performed with the earliest automation hardware, the Model 12 Auto Ion Analyzer, but they can certainly be performed more conveniently with more modern hardware. Likewise, the Model 12 automated ion chromatograph was used in a different study to determine bromide by electrochemical detection.[47] A silver electrode was used as the working electrode with a platinum counter electrode and a Ag/AgCl reference electrode. Bromide levels as low as 0.03 mg/$\ell$ were detected.

Other trace ions of environmental concern have been determined in water samples. Ion chromatography was used together with an atomic absorption spectrometer (AAS) to separate and detect organic and inorganic species.[48] The original AS-1 column was used with the standard 3 m$M$ sodium bicarbonate/2.4 m$M$ sodium carbonate eluent to separate monomethyl arsonic acid (MMA) and dimethyl arsinic acid (DMA). These anions have low specific conductance, which limited the use of chemically suppressed conductivity detection. Instead, the MMA and DMA were converted to arsine-generating system interfaced to the AAS that was first described by Vijan and Wood.[49] Arsine derivatives are formed by reaction with sodium borohydride. The generated arsine derivatives could then be detected by AAS. The large concentration of common ions such as chloride do not interfere with the AAS, whereas they do interfere with chemically suppressed conductivity. As a result, as many as five different arsenic-containing species can be selectively detected. These include DMA, MMA, arsenite, arsenate, and $p$-aminophenyl arsonic acid. This method was used to measure the concentrations of organo-arsenicals in air samples, though the method can be applied to water samples.

Not only arsenic, but also selenium compounds can be determined using ion chromatography. Although the organo-arsenics are not detectable at low levels using chemically suppressed conductivity, the inorganics, arsenite, arsenate, along with selenite and selenate, are. A separation of these from six other anions that might exist in water samples is shown in Figure 22. Another important environmentally hazardous metal is Cr(VI), which exists as the anion, chromate. The most sensitive detector for chromate is a UV-visible detector. Asshown in Figure 23, as little as 0.01 mg/$\ell$ (10 ppb) can be easily detected. It should be noted that an MPIC guard column is used in this analysis. It removes organics from the sample that could eventually poison an anion-separator column.

The electrical utility industry has also found ion chromatography to be very useful in analyzing their water samples. Some ions, such as chloride, are particularly corrosive, causing Chemistry Guidelines to be written by the Steam Generator Owners Group, and the BWR Owners Group.[50] As a result, water samples at the Sequoyah Nuclear Plant in the Tennessee Valley Authority (TVA) were analyzed.[51] These water samples were analyzed automatically using a four-channel, 18-stream, computer-controlled Series 8000® ion chromatograph. The steam samples analyzed were (1) combined steam generator blowdown, (2) main steam, (3) hot well, (4) condensate, and (5) feed water. Dilution modules were used to dilute more stable, concentrated standards down to the ppb-range found in the steam samples. Six cations and seven anions were determined. Using this system, detection-limits as low as 20 parts per trillion (ppt) were reported for choride and 50 ppt for sulfate. These represent concentrations of the anions in the original, unconcentrated steams. It is important for electric utilities to detect such low levels, because very large volumes of steam are used each day. If water contains as much as 1 μg of highly corrosive chloride per liter, turbine blades may be exposed to as much as 1 kg of chloride in continuous use.

Ion chromatography is also useful in analyzing heavily contaminated and deeply colored water samples such as those found at the fixed-bed coal gasifier at the University of North

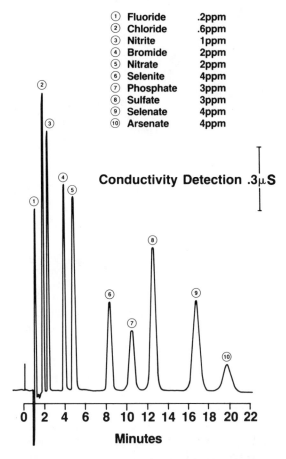

| | | |
|---|---|---|
| ① | Fluoride | .2ppm |
| ② | Chloride | .6ppm |
| ③ | Nitrite | 1ppm |
| ④ | Bromide | 2ppm |
| ⑤ | Nitrate | 2ppm |
| ⑥ | Selenite | 4ppm |
| ⑦ | Phosphate | 3ppm |
| ⑧ | Sulfate | 3ppm |
| ⑨ | Selenate | 4ppm |
| ⑩ | Arsenate | 4ppm |

Conductivity Detection .3μS

Minutes

FIGURE 22.    Arsenic, selenium, and common anions. AG-5 guard column; AS-4A separator column.

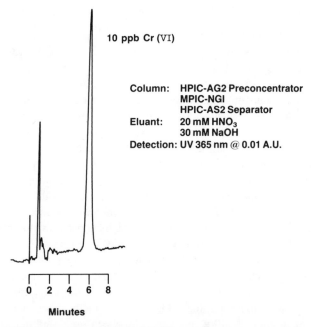

10 ppb Cr (VI)

Column:    HPIC-AG2 Preconcentrator
           MPIC-NGI
           HPIC-AS2 Separator
Eluant:    20 mM HNO₃
           30 mM NaOH
Detection: UV 365 nm @ 0.01 A.U.

Minutes

FIGURE 23.    Determination of Cr(VI) in drinking water, at low ppb levels.

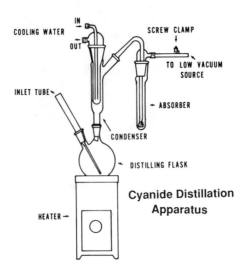

FIGURE 24.    Cyanide sample preparation.

Dakota Energy Research Center.[52] These waters were analyzed for nitrate, nitrite, phosphate, sulfate, thiocyanate, and thiosulfate. An AS-3 column was used for separating the common anions. The large, hydrophobic thiocyanate and thiosulfate are strongly retained on the AS-3, requiring the use of the AS-5 column and an eluent containing 4.5 m$M$ sodium bicarbonate, 3.6 m$M$ sodium carbonate, 3.36 m$M$ p-cyanophenol, and 2% acetonitrile. Unlike the highly purified waters in steam samples, these samples had high concentrations of anions and required no preconcentration.

Another ion of significant environmental impact is cyanide. Total cyanide in liquids, solids, and sludges can be determined by first distilling the cyanide out of the sample with an apparatus such as that shown in Figure 24. The sample is boiled in an acid solution, and the volatile HCN is condensed into a collection tube containing a strongly alkaline solution. The cyanide is separated from other anions and is then detected using a silver electrode in an electrochemical detector (see Chapter 3, Section V for details of the procedure). Sample chromatograms illustrating the separation of cyanide from sulfide (which interferes in other amperometric methods) is shown in Figure 25. As little as 50 ppb cyanide were detected with a relative standard deviation of 6.4%.[53] Water samples analyzed, included treated organic cleaner, waste-treatment sludge, landfill leachate, wastewater, treated pickle liquor, scrubber water, and metal-finishing water. When analyzing metal-finishing wastewater, it is possible to separate metal cyanides by MPIC. In addition to the seven different types of samples just mentioned that were found to have low (parts per billion) levels of cyanide, some 12 other samples containing higher levels of cyanide were analyzed. The low level precision (standard deviation) is 3 ppb as compared to 10 ppb for the standard colorimetric method.[54] The bias of the ion-chromatographic method relative to the colorimetric method is −13%. Thus, cyanide can be determined in highly alkaline solutions such as those used in total cyanide analysis by ion chromatography with electrochemical detection.

Another aqueous sample that has been analyzed by ion chromatography is fog water. Bisulfite ($HSO_3^-$), the aqueous form of sulfur dioxide, forms an adduct with formaldehyde in fog and cloud water to produce hydroxymethanesulfonate.[55] Bakersfield, Calif. is one location where sufficient atmospheric sulfur dioxide exists to be measured. Dynamic models of fog and cloud water chemistry[56] indicate that HMSA could be the major source of S(IV) at pH above 5.[57] It was suggested that a likely environment for rapid HMSA formation is in plumes that are sources of sulfur dioxide. It has been suggested that S(IV) and formal-

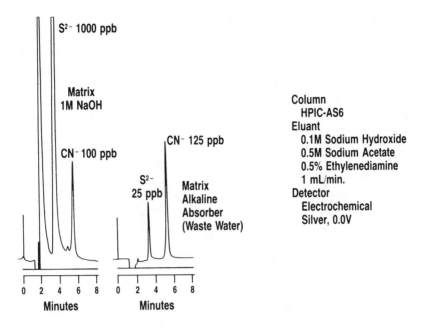

FIGURE 25.   Cyanide in strongly alkaline solutions. Column: AS-6, electrochemical detection.

dehyde in aerosol are supersaturated in Los Angeles[58,59] and Germany.[60,61] Total concentrations of HMSA ranged from 0 to 140 $\mu M$ in 1984 and 1985. HMSA was separated from other anions using MPIC and ion pairing. The ion-pairing reagent used is 2 m$M$ tetrabutyl ammonium chloride. The organic modifier was 5% methanol. Detection was accomplished by using a suppressor in the silver form (i.e., the original packed-bed ICE suppressor). Thus, ion chromatography has been used to determine much more than common anions and cations in water samples.

Water is not the only substance that is of environmental concern. Ion chromatography can also be used to monitor air pollution. Coal fly-ash, high- and low-altitude atmospheric samples, roadside samples, and gas streams can all be analyzed. In fact, the first major applications for ion chromatography were aerosol- and airborne-particulate-matter samples. The majority of papers presented in 1977 and 1978 at the Symposia on Ion Chromatographic Analysis of Environmental Pollutants discussed the analysis of aerosols, particulates, or other atmospheric samples. In every case, ingenious sample-collection devices were used.

In the case of sulfur dioxide, the sampling unit used in one of the earliest reports[62] consisted of a restricted-orifice bubbler immersed in a peroxide-absorbing solution, a trap to remove moisture, a critical orifice, and a vacuum pump to pull air at about 200 m$\ell$/min. The sulfur dioxide is converted to sulfate in the peroxide solution. The sulfate content of the peroxide solution is then measured by ion chromatography. The accuracy of this method was tested by a sulfur-dioxide test atmosphere generator. Known amounts of sulfur dioxide were generated, and the percent recovery of the sulfur dioxide (as sulfate) in the peroxide-absorbing solution was determined to be 100% (values ranged from 94 to 116%). Atmospheric sulfur-dioxide concentrations in ambient air at the EPA-annex parking lot, ranged from 26.0 to 31.7 $\mu$g/m$^3$ of air.

The other major sulfur-based air pollutant is sulfuric acid, which is produced when sulfur-containing coal is burned in the generation of electric energy. In addition there are significant amounts of sulfuric-acid aerosols generated by automobile catalytic converters. Sulfuric acid aerosols can be collected on Teflon® filters, but other atmospheric pollutants were shown to react with sulfuric acid on the filter, producing a negative bias in the results.[63] As a result, it is necessary to stabilize the sulfuric acid during sampling. This is done by removing interfering pollutants in a series of traps. The sulfuric acid is then selectively volatilized and is then prefiltered and collected by derivatization on an alkali-impregnated Teflon® filter. The sampling apparatus consisted of a 3.2-mm inner diameter stainless-steel tube through which the ambient air enters. The sample of air is then passed through a cyclone which removes all airborne-particulate matter larger than 2 μm in diameter. The air sample is then passed through a dehumidifier and two gas-diffusion denuders, which remove ammonia and sulfur dioxide. The first denuder is coated with phosphoric acid for removing alkaline gases such as ammonia. The second denuder is coated with sodium hydroxide for removing acidic gases such as sulfur dioxide. The dehumidifier prevents the hydroscopic phosphoric acid and sodium hydroxide-coated denuders from building up too much moisture. After passing through the denuders, the air stream is heated to 130°C in a Teflon®-coated heating zone. The selectively volatilized sulfuric acid is then collected on an alkali-impregnated Teflon® filter. This prototype sampler was set up in a parking deck adjacent to an elevated (8 m) six-lane interstate highway north of Birmingham, Ala. The values of sulfuric acid measured varied from 0.5 to 3 μg/m³. Thus, with proper sampling, ion chromatography can measure both sulfur dioxide and sulfuric acid in atmospheric samples.

This is not the only sampling method used however. Other workers have used charcoal sorbent for collecting sulfur dioxide[64] or Teflon® filters for determining total anions in air.[65] Detection limits as low as 5 ng chloride per cubic meter of air were reported. It should be emphasized that these studies all used the AS-1 separator column. Certainly the ion-chromatographic results are valid, even though the anions can now be separated faster using separator columns such as the AS-4A.

The other major type of common atmospheric sample is atmospheric-particulate material (APM). APM is usually collected on a filter. The filter can then be extracted with an aqueous solution to extract the anions that have been trapped. The anions are then quantitated by ion chromatography. In one study, APM was collected on Fluoropore® FA membrane filters, and the filters were extracted with 0.05 m$M$ perchloric acid. Aliquots of the extracts were analyzed by ion chromatography and by an automated wet chemical method. The results obtained with wet-chemical analysis were plotted against the results obtained by ion chromatography. The data for nitrate analysis showed a correlation coefficient of 0.995, and for sulfate, 0.9995. Thus, the two methods found similar concentrations in the samples analyzed.

Just as the concentration of sulfur compounds in atmospheric samples is important, so also is the concentration in the source of the pollution, i.e., fuel oils and coal. The sulfur in these organic samples is usually present as covalently bound sulfur, which cannot be simply extracted into an aqueous solution. Instead, the covalent bond must be broken by pyrohydrolysis, or combustion. A Parr bomb or Schoniger flask can be used for this task. In one study, a Parr bomb was used, and the sulfate formed by sample combustion was determined by ion chromatography and by titrimetry.[66] The results obtained by the two methods showed good agreement. Similar comparisons have been made not just in coal analysis, but also fuel oils and gasoline samples. The results, summarized in Tables 2 to 4, indicate that the ion-chromatographic method agrees well with standard (ASTM) methods.

Ion chromatography is also useful in analyzing air and water samples of importance in industrial hygiene. Airborne inorganic acids can be collected on a silica gel absorber, and be determined as their anions, fluoride, chloride, phosphate, nitrate, and sulfate. Hydrogen cyanide in air was determined by converting the cyanide to formate by using a 0.2 $N$ NaOH-absorbing solution.[67] The formate was separated from other anions using an AS-1 separator and a weak 5 m$M$ sodium tetraborate eluent. Formate eluted in about 9 min when the flow-rate was 138 mℓ/hr. Another source of cyanide is in the reprocessing of cathodes in the aluminum industry. An ion-chromatographic method for determining cyanide in these samples has appeared.[54] It is estimated that as much as 121,000 kg of cyanide is released by

## Table 2
## SULFUR IN COAL

| | Sulfur, (%) | |
|---|---|---|
| Sample no. | IC | ASTM D-1552 |
| 1 | 2.49 | 2.58 |
| 2 | 2.59 | 2.57 |
| 3 | 2.42 | 2.57 |
| 4 | 2.54 | 2.64 |
| 5 | 2.55 | 2.64 |
| 6 | 2.49 | 2.57 |
| 7 | 2.45 | 2.55 |
| 8 | 2.54 | 2.52 |
| 9 | 2.49 | 2.60 |
| 10 | 2.49 | 2.52 |
| Average | 2.51 ± 0.050 | 2.58 ± 0.042 |
| % RSD | 2 | 2 |

## Table 3
## SULFUR IN FUEL AND DIESEL OILS

| | Sulfur, (%) | |
|---|---|---|
| Sample no. | IC | ASTM-129 |
| F-77-002-242 | 0.202 | 0.209 |
| F-77-002-243 | 0.219 | 0.229 |
| F-77-002-244 | 0.268 | 0.238 |
| F-77-002-245 | 0.257 | 0.258 |
| F-77-002-246 | 0.203 | 0.206 |
| F-77-002-247 | 0.084 | 0.087 |
| F-77-002-248 | 0.094 | 0.093 |
| F-77-002-249 | 0.166 | 0.176 |
| F-77-002-257 | 0.095 | 0.118 |
| F-77-002-258 | 0.175 | 0.171 |
| F-77-002-259 | 0.267 | 0.253 |
| F-77-002-260 | 0.197 | 0.198 |
| F-77-002-261 | 0.260 | 0.234 |
| F-77-002-262 | 0.152 | 0.179 |
| F-77-002-263 | 0.198 | 0.195 |
| NBS 1624 | 0.211 | 0.209 |

## Table 4
## SULFUR IN GASOLINES

| | Sulfur, (%) | |
|---|---|---|
| Sample no. | IC | ASTM D-1266 |
| F-78-000-801 | 0.027 | 0.025 |
| F-78-000-802 | 0.036 | 0.036, 0.035 |
| F-78-000-803 | 0.021 | 0.015, 0.022 |
| F-78-000-805 | 0.014, 0.013 | 0.009, 0.011 |
| F-78-000-806 | 0.019, 0.019 | 0.014 |
| F-78-000-807 | 0.012, 0.012 | 0.017 |
| F-78-000-808 | 0.025 | 0.026 |
| F-78-000-809 | 0.018 | 0.021 |

**Table 5**
**RESULTS FOR VALIDATION OF**
**CAC METHOD**

| CAC Air concentration (ppm)[a] | Expected chloroacetate conc. ($\mu$g/m$\ell$)[b] | Found chloroacetate conc. ($\mu$g/m$\ell$)[c] |
|---|---|---|
| 0.83 | 31.6 | 30.0 (0.013) |
| 0.77 | 29.5 | 29.2 (0.029) |
| 0.42 | 15.8 | 16.1 (0.016) |
| 0.39 | 14.8 | 15.2 (0.033) |
| 0.17 | 6.3 | 6.3 (0.063) |
| 0.15 | 5.9 | 6.4 (0.039) |
| 0.083 | 3.2 | 3.1 (0.032) |
| 0.077 | 3.0 | 3.1 (0.032) |
| 0.042 | 1.6 | 1.6 (0.156) |
| 0.039 | 1.5 | 1.5 (0.133) |

[a]  Based on a 20-liter air sample.
[b]  Based on weight losses from CAC diffusion tubes used.
[c]  Results presented as average of three samples, followed by the relative standard deviation (RSD) in parentheses.

the aluminum industry each year. Samples considered typical of aluminum wastewaters that contained not just cyanide, but also suspended solids, trace elements, and other substances were taken. Cyanide and complex cyanide were converted to HCN by refluxing the sample in a mineral acid with magnesium ion present, as specified by the EPA.[68] The samples were then removed from the scrubbing vessel and analyzed by ion chromatography and spectrophotometry; 56 samples and 10 standards were analyzed by both methods. The ion chromatographic results were plotted against the spectrophotometric results, and a correlation coefficient of 0.991 was obtained, suggesting that similar results are obtained with the two methods.

Atmospheric formaldehyde and acetaldehyde were determined in a separate study by using an alkaline hydrogen peroxide impinger system.[69] This system oxidizes the aldehyde to formic acid and acetic acid which were separated on an AS-1 using a 1.5 m$M$ sodium bicarbonate eluent. Similarly, sulfuryl fluoride, which is used as a fumigant for termite control was collected on a charcoal filter.[70] The sulfuryl fluoride was then hydrolyzed to fluoride and sulfate, which can be quantitated by ion chromatography. One of the more important analyses in industrial hygiene is chloroacetyl chloride (CAC), which is a sensory irritant to vinyl chloride. It is collected on silica gel, desorbed with water, then injected on an anion-separator column. Using an AS-1 and a 1.5 m$M$-sodium bicarbonate eluent, CAC elutes in about 5 min and is well separated from chloride and other anions typically present. The results of repetitive analyses of standards are summarized in Table 5. Ion chromatography has also been used to determine nitrites in metal-cutting fluids.[71] Nitrites are added to cutting fluids to inhibit rust formation. To optimize accuracy in determining nitrite, it is important to use either a fiber suppressor or a micromembrane suppressor. Another application is the determination of azide in automobile air bags. The air bags inflate when sodium azide is acidified, producing nitrogen gas. Azide elutes between nitrite and nitrate on an anion separator and is detected by chemically suppressed conductivity as shown in Figure 26. Another pollutant, ethanolamines, can be determined using MPIC. Ion pairs are formed between the ethanolamines and octanesulfonic acid.[72] The original report suggested using a

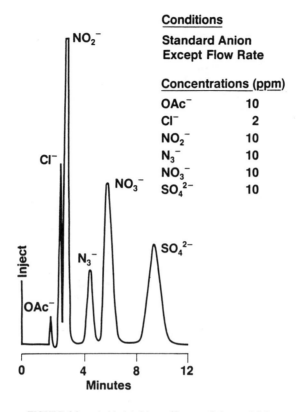

FIGURE 26.    Azide in airbag effluents. Column: AS-3.

packed-bed cation-suppressor column. More recently, a continuously regenerated micro-membrane suppressor can be used for MPIC. Ethanolamines are used to scrub carbon dioxide and hydrogen sulfide in refinery water. Thus, ion chromatography is quite useful in determining a number of compounds and ions of environmental concern. The applications are summarized in Table 6.

## III. PLATING SOLUTION ANALYSIS

The emergence of ion chromatography has been a significant aid to the electronics and electroplating industries. The manufacture of many electronic components, such as hybrid microcircuits, requires the use of many different plating solutions. Electroplating also has significant applications in metal processing for products as varied as automobiles and industrial construction. Quite often, the plating solutions contain a complex mixture of ionic and nonionic components. They quite often contain one or two major chemicals, and the analyst is required to determine not only these, but also the numerous minor constitutents. They are often quite corrosive, varying from concentrated chromic acid to hot, concentrated alkali. The nature of these samples demands the use of polystyrene/divinylbenzene resins that are stable from pH 0 to 14, and nonmetallic, corrosion-resistant hardware present in an ion chromatograph. Before the emergence of ion chromatography, there were often no methods available to accurately determine trace ions in the presence of a large excess of a primary ion or ions. For example, the determination of ppm-levels of chloride in sulfuric-acid-anodizing baths would require precipitation of the sulfate, a tedious procedure with questionable reliability. Even nonionic components such as detergents in an alkaline cleaning bath could not be determined by conventional HPLC because the silica-based columns are

**Table 6**
**SUMMARY OF ENVIRONMENTAL APPLICATIONS**

| Sample | Analytes | Separation mode | Detection mode |
|---|---|---|---|
| Air | Sulfate, chloride, nitrate, phosphate, azide, and chloroacetyl chloride | Anion exchange | Conductivity |
| Air | Formaldehyde | ICE | Conductivity |
| Air | Cyanide | Anion exchange | Electrochemical |
| Acid rain | Sulfate, chloride, and nitrate | Anion exchange | Conductivity |
| Snow and ice | Sulfate, chloride and nitrate | Anion exchange | Conductivity |
| Snow and ice | Monovalent cations | Cation exchange | Conductivity |
| Drinking water | Seven common anions | Anion exchange | Conductivity |
| Drinking water | Mono- and divalent cations | Cation exchange | Conductivity |
| Drinking water | Transition metals | Cation exchange | Post-column reaction, visible detection |
| Forest canopy | Calcium | Cation exchange | Conductivity |
| Forest canopy | Sulfate and nitrate | Anion exchange | Conductivity |
| Industrial waters | Seven common anions and sulfite | Anion exchange | Conductivity |
| Industrial waters | Mono- and divalent cations | Cation exchange | Conductivity |
| Industrial waters | Lanthanide and transition metals | Cation exchange | Post-column reaction, visible detection |
| Industrial waters | Ethanolamines | MPIC | Conductivity |
| Industrial waters | Chromium, arsenic, and selenium | Anion exchange of oxy-acid anions | UV or conductivity |
| Industrial waters | Thiocyanate, thiosulfate | Anion exchange | Conductivity |

not stable to high pH. Spectroscopic methods such as infrared, mass spectroscopy, and even NMR are not amenable to either the water, or the high ionic strength of the plating solutions. In many analyses, ion chromatography provides the only possible analytical method.

Of special interest to the industrial chemist analyzing plating solutions, is the unique combination of versatility and automation provided by the number of separators and detectors and microprocessor control. This enables fully unattended analysis of plating baths for anions, cations, and even nonionic components. To best utilize this capability, however, it is necessary to understand which separators, eluents, and detectors are needed for each type of analysis and then plumb the ion chromatograph in such a way as to optimize versatility and automation.

Probably the most obvious application for plating-solution analysis is the determination of the metals being electroplated. The one method most extensively discussed in the literature is transition metal determinations. The CS-2 column was used for separation, and post-column reaction with PAR for detection.[74-77] As discussed previously (Chapter 2, Section VIII), the relative-retention times (or capacity factor, k′) of the transition metals can be adjusted by making appropriate changes in eluent composition and pH. A 10-m$M$ oxalic acid/7.5-m$M$ citric acid eluent at pH 4.3 with LiOH was originally used for obtaining a profile of the most common transition metals.[74] Ferrous and ferric ions are simultaneously determined along with at least five other metals. Some metals, such as cadmium and manganese, are tightly bound by the CS-2 column and require a stronger eluent, so 40 m$M$ tartaric acid with 12 m$M$ citric acid is used.

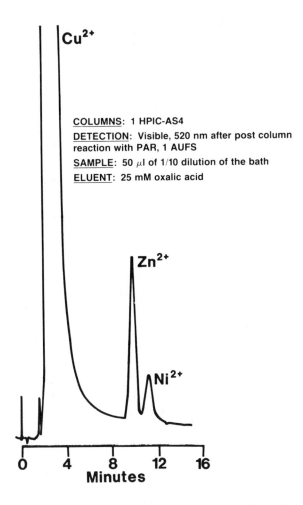

COLUMNS: 1 HPIC-AS4
DETECTION: Visible, 520 nm after post column reaction with PAR, 1 AUFS
SAMPLE: 50 $\mu$l of 1/10 dilution of the bath
ELUENT: 25 mM oxalic acid

FIGURE 27. Metal contaminants in an acid copper plating bath. Column: CS-2; Detection: post-column reaction with PAR and visible detection.

The plating analyst is often interested in a seemingly less complex method, however. It is very important to monitor the depletion of metal from the bath, making rapid, more reliable determination of the one or two major metals essential. For example, if separate zinc, copper, nickel, cobalt, and lead baths are to be analyzed, it is quite possible that separation is not important because only one major component is present. In this case, it might be best to use the strongest eluent possible. With an eluent containing higher citric acid and tartaric acid than 12 and 40 m$M$ could potentially permit analysis of samples in 5 min, enabling 100 samples to be analyzed in an unattended mode within 500 min. All that would be required would be to have a technologist or a robot arm dilute samples and load them into the autosampler. The attached computer or computing integrator can make all calibration calculations and print the results in report format. In fact, an instrument now exists, the Series 8000®, than can even automate the sample acquisition, dilutions, and loading into the autosampler.

Ion chromatography is not quite limited to determining only the major metal in a plating bath. The same CS-2 column with post-column reaction with PAR can be used to determine trace metals. As seen in Figure 27, low ppm-levels of zinc and nickel can be determined in an acid copper bath. Also, trace levels of lead, copper, cadmium, and cobalt can be

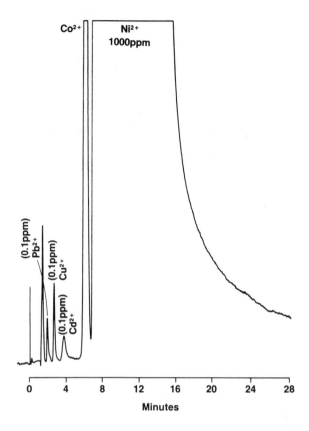

FIGURE 28.    Trace metals in electroless nickel plating bath.
Column: HPIC-CS-5.

determined in an electroless nickel bath as shown in Figure 28. There are several other examples of metals determinations in the literature. Some of these include the determinations of metal contaminants in an acid etchant, an acid copper bath, and in steel etchants.[75,76] Zinc, cobalt, and lead were determined in the presence of an excess of Fe in a steel etchant,[75] zinc and nickel were determined in a copper sulfate plating bath. The cobalt cyanide complex was determined in a gold cyanide bath.[76]

Since these reports appeared, however, a new column, the CS-5, has been developed. This column has a higher efficiency (about 12,000 plates per meter). It is packed with a 13-$\mu$m polystyrene/divinylbenzene substrate resin agglomerated with a special latex to provide this efficiency.

Transition metals can, paradoxically, be determined using anion exchange by two approaches. In the first approach, EDTA is added to form an anionic metal-EDTA complex.[77,78] In some cases, samples can be prepared by simply mixing 1 m$\ell$ of 0.025 $M$ disodium EDTA with a sample containing less than 50 ppm of total metals. Of course, other anions present will also be detected, making this a method for simultaneous determination of anions and cations. An AS-5 column is recommended for this analysis, because excess unreacted EDTA is strongly retained on other anion separators. It could affect the utility of an AS-1, 2, 3, 4 or 4A for other types of analyses because the EDTA would not be completely washed off the separator. As with most other anions, the free EDTA is not retained as strongly on the AS-5 column and can be washed off sooner, allowing the column to be used for other analyses.

There is another approach to using anion-exchange columns to separate transition metals.

Anion-separator columns have some free sulfonate groups to which the quaternary amine-containing latex beads did not bind. These free-sulfonate groups are cation-exchange sites. Thus, columns such as the AS-1 and 4 will bind small amounts of transition metals and have been reported to be used to separate them. Haak and Franklin[76] reported the determination of zinc and nickel contaminants in an acid copper-plating bath using an AS-4 column with 25 m$M$ oxalic acid eluent followed by post-column reaction with PAR.

Thus, there have been several methods reported for determining transition metals in plating solutions. The majority of these use the CS-2 column and either oxalate/citrate or tartarate/citrate eluents. Certainly these reports are valid, and the analyses easily reproduced. On the other hand, transition metal analysis provides another example where technology is ahead of the chemical literature. The fastest, most convenient method for transition metal analysis uses the CS-5 column, which was not available when earlier reports appeared.[74-77] In fact, even more methods using the CS-2 are likely to be reported at future conferences and in the electroplating industry.

Although transition-metal analysis may be the most obvious application for plating chemists, many other applications exist. In the case of rapid, simultaneous determination of anions, especially in highly corrosive baths, no other method can compare. Perhaps it is for this reason that the majority of methods in the literature are for anion analysis. In fact, many plating solutions contain anions, especially chloride, whose concentrations need to be controlled. Alkaline cleaners need to be analyzed for phosphate, acid zinc baths for total chloride, and low conductivity deionized water rinses for total anion concentrations. In the case of alkaline and acidic baths, corrosion-resistant columns, pumps, and other hardware are quite useful. If compatible analyses are properly planned, quite diverse plating solutions could be analyzed for different anions. Thus, baths which do not contain large, hydrophobic anions (such as chromate) can be analyzed on an AS-4 using a bicarbonate/carbonate eluent. Baths that do contain large, hydrophobic anions could be analyzed by automatically switching to an AS-5 column, and analysis could proceed using either the same bicarbonate/carbonate or another eluent.

The analysis of many plating baths is trivial. They need only be diluted, filtered, and injected on an AS-1, 2, 3, 4, or 4A column using an isocratic bicarbonate/carbonate eluent and chemically suppressed conductivity detection. Examples include chloride and sulfate in deionized water rinses; total chloride in zinc chloride/HCl baths to help monitor total acidity; chloride, sulfamate, and sulfate in a nickel sulfamate bath; fluoroborate in a tin/lead fluoroborate plus fluoroboric acid bath; hypophosphite, phosphite, chloride, phosphate, and sulfate in an electroless nickel bath formate sulfate, and tartarate in an electroless copper bath; citrate, phosphate, and nitrate in a different electroless copper bath; and flouride and chloride an an ammonium bifluoride/HCl etchant.[75-77] In each of these cases, no one anion is present at a concentration so high that it would not affect the chromatographic properties of the anions. To perform all these analyses, it would be necessary to first analyze a set of standards containing the anions of interest. The computer can fit this calibration data to a straight line. Although the analytical methods development may seem trivial, the data provided to the electroplater is not. Take, for example, the electroless nickel bath. The current needed to reduce $Ni^{2+}$ to nickel metal is provided chemically by the strong reducing agent, hypophosphite. As the nickel is electroplated (i.e., reduced), the hypophosphite is oxidized to phosphite. The phosphite can also be oxidized to phosphate. Thus, to accurately monitor bath performance, it is important to know that concentration of all three phosphorous species. With ion chromatography this is easy, as seen in Figure 29. By providing simultaneous, multicomponent analysis, the analytical chemist is able to greatly assist the electroplater.

In many cases, the analyst is required to determine trace anions in the presence of a large excess of other anions. Examples include chloride in a sulfuric-acid-anodizing bath, and chloride in sodium hydroxide. In general, it is best to dilute the sample as much as possible

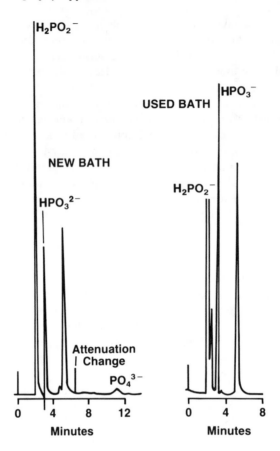

FIGURE 29.   Determination of hypophosphite, phosphite, and phosphate in electroless nickel baths.

to avoid column overloading due to the major ionic constituent. Unfortunately, one may become accustomed to performing more routine analyses differently. When determining anions that are all present at roughly the same concentrations (such as anions in an electroless nickel bath), it is probably best to dilute the bath to such an extent so as to allow usage of the 10 μS full-scale conductivity-detector setting. At this scale, the anion separator would not be overloaded and the signal being measured would be large enough to permit good precision (relative standard deviations about 2%). On a 10-μS scale, the water dip would be minimal. However, if one dilutes a sulfuric-acid-anodizing bath only enough to detect a sizable chloride peak at 10 μS, the sulfate peak would be so large as to completely mask the chloride, and the column would be seriously overloaded. Thus it is better to dilute the sample another tenfold and use the 1-μS scale. It is also important to prepare standards that mimic the sample, by containing the same amount of major ions that are known to be in the samples.

In the case of trace anion determinations in sodium hydroxide, the introduction of the micromembrane suppressor provided a significant improvement. In an earlier report,[79] the packed-bed suppressor was used and a large negative peak (the ''water dip'') appeared in the published chromatograms. The advent of the anion-fiber suppressor helped because the water dip did not change in size or in elution time, quite unlike results obtained from a packed-bed suppressor. The fiber suppressor did have a limited capacity for suppression. A large negative peak arose from the large excess of unsuppressed hydroxide in the NaOH. The fiber suppressor was able to convert the hydroxide to water, but only in limited amounts.

If determining large amounts of chloride in technical grade NaOH, it is only necessary to sufficiently dilute the NaOH to keep from overloading the suppressor. The anion micro-membrane suppressor, on the other hand, can suppress far more OH than either the packed-bed or the fiber suppressor. Thus, more concentrated samples of NaOH can now be injected on an anion-separator column and only small water dips produced. To minimize matrix effects, NaOH itself is used in the eluent. An anion separator with a very high anion exchange capacity, the AS-6, was used. The eluent also contains 20 m$M$ 4-cyanophenate, which can elute analytes within 20 min and can be converted in the suppressor to 4-cyanophenol, which has very low conductivity. The presence of NaOH ensures the ionization of the 4-cyano-phenol, it minimizes the jump in pH upon sample injection, and it fully ionizes the phosphate in the sample, increasing its retention. It should be remembered that a given increase in eluent concentration affects the retention times of divalent anions more so than monovalents. In this way, the concentrations of the NaOH and 4-cyanophenate could be altered to produce optimum separation of the divalents, carbonate, sulfate, oxalate, and sulfate. The water dip is still sizable, because large amounts of hydroxide in the sample are being converted to water by the suppressor. Because water has a much lower conductivity than eluent (even after suppression), a water dip results.

The maximum concentration of NaOH sample that can be loaded onto a concentrator column without eluting weakly retained carbonate and chloride, was determined.[80] Successive dilutions of a spiked sample were analyzed. The plot of detector response vs. NaOH concentrations was linear up to 1% NaOH by weight. The optimum sample concentration for routine analysis was reported to be 0.5% for high-purity NaOH, and 0.05% for unpurified NaOH. The extra dilution to 0.05% was required to keep the large carbonate and chloride from overloading the column. Samples that were originally presented as 50% NaOH were found to have detection limits on the order of 15 to 80 ppb for chloride and sulfate. Calibrations were linear through 50 ppm chloride and phosphate. Typical concentrations in real NaOH samples were reported to range from 7.5 to 24 ppm for chloride, and 6.6 to 9.8 ppm for phosphate. Nitrite, sulfate, oxalate, chlorate, and nitrate were also found, but at sub-ppm levels.[80]

A special problem can arise when trying to determine trace chloride in plating baths containing large amounts of a major anion. Many such plating baths contain organic additives which affect grain size and brightness of the electrodeposit. These brighteners often contain unknown organics that can interfere by co-eluting with chloride. To separate the chloride from these organics, it may be necessary to use a weak eluent such as sodium bicarbonate (without sodium carbonate) or sodium borate.[3] With these eluents, however, nitrate and sulfate are strongly retained on the AS-1, 2, 3, 4, and 4A columns. To elute the nitrate, sulfate, and any other divalent anions, it is necessary to switch back to a stronger eluent. Thus, a step gradient from 4 m$M$ sodium carbonate to 3.0 m$M$ sodium bicarbonate plus 2.4 m$M$ sodium carbonate in determining trace chloride in acid copper-sulfate baths. The eluent is then switched back to the 4 m$M$ sodium bicarbonate to reequilibrate the column before the next injection. Again, this can all be done automatically using microprocessor or computer controls.

It should be emphasized that even though earlier reports appeared, which showed the separation of inorganic anions in plating baths using an AS-1 or AS-3 column, it is quite likely that the analyses could now be performed faster on an AS-4 or AS-4A. As stated previously (Chapter 2, Section I) the AS-4A is quite similar to the AS-4, but is more rugged, with special utility for samples which contain components that may poison an AS-4. Although the AS-4A was originally designed for the analysis of soil samples, plating baths can present equally complex matrices. Quite often, the plating chemist is not sure of the exact chemical composition, especially if organics are present. Thus, it is probably better to use the rugged AS-4A column whenever possible when determining anions in plating baths.

In some cases, the AS-4A should not be used. Some baths contain large, hydrophobic anions that are strongly retained on the AS-4A (and the AS-1, 2, 3, and 4). One example is chromate. Often chromic acid and potassium dichromate baths must have their chloride and sulfate content monitored. To do this, an aliquot of the bath is neutralized with either sodium carbonate or sodium bicarbonate, and is diluted into the 3.0-m$M$ sodium bicarbonate/ 2.4 m$M$ sodium carbonate eluent to be used. The chloride and sulfate elute earlier on the AS-5 than on the AS-4A column. More importantly, the chromate would be strongly retained on the AS-4A using this eluent. Some reports have appeared which suggest using a guard column, i.e., a short separator, and a strong sodium chloride eluent to elute chromate along with the chloride and sulfate.[81] These reports were written before the advent of the AS-5 column. The AS-5 provides much better resolution than the guard column and can be used for analyzing other plating baths as well. Thus, it is very convenient to connect the AS-5 and AS-4A to the same valve (the A valve) on the ion chromatograph, and use microprocessor or computer control to automatically select the desired column for each anaysis.

The output from this valve can be directed to the one suppressor column (e.g., an anion micromembrane suppressor). Thus, effluent off the AS-4A or AS-5 column would be directed to the same suppressor. This is especially convenient because the same eluent can be used for both the AS-4A and AS-5 and very little instument time would be used to equilibrate the system when switching columns. Thus, one could load standards into the autosampler along with diluted plating-bath samples for analysis on the AS-4A (examples might include total chloride in a zinc chloride/HCl bath, fluoride and nitrate in an ammonium bifluoride/ nitric acid bath, and anions in deionized water rinses). The concentrations of each component in the standard are entered into the computer, and the calibration data fit to a straight line. This calibration data is used, together with the dilution factors of the samples, to automatically calculate the concentrations of the anions in the baths. After this is done, the computer can switch the A valve to select the AS-5 column. After about 15 min of equilibration with the eluent, appropriate standards for the AS-5 can be analyzed. The samples containing large, hydrophobic anions can then be analyzed, and the concentrations automatically calculated, based on the calibration data.

The AS-5 is very useful for other analyses, too. It can be used for simultaneous anion and metal analysis if the metals are first chelated with EDTA to form an anionic metal-EDTA complex. Thus, the metals are actually separated as anions. Because free EDTA is strongly retained on other anion separator columns, the AS-5 should be used. The EDTA is eluted as a broad peak off the AS-5. Tin/lead alloy baths can be analyzed simultaneously for lead and fluoroborate by simply adding EDTA to the sample.[82] The Pb-EDTA complex elutes in about 1.7 min, and is well separated from the fluoroborate, which elutes shortly after sulfate on the AS-5. Other plating baths can be analyzed for other transition metals (and anions) in this way. An important consideration, however, is to use a Chelex, or metal-free, column to clean any metal contamination from the eluent. The Chelex column is a high-capacity column which binds cations that would form complexes with EDTA. If the Chelex column was not used, spurious peaks can arise due to EDTA complexes with such contaminants (like calcium and magnesium) in the eluent.

One application using a metal-EDTA complex requires special sample treatment. When using concentrated chromic acid to strip epoxy or other adhesive off printed wiring board, it is often important to know when too much chromate has been reduced to $Cr^{3+}$, and the stripping solution rendered ineffective. As mentioned previously, $Cr^{3+}$ and chromate can be determined simultaneously by separation on a CS-5 followed by UV-visible detection. This method requires heating the sample in a pyridine dicarboxylic acid (PDCA) solution to form a Cr-PDCA complex. If the concentration of chromate is too high, this presents a problem. If concentrated chromic acid were heated with PDCA, the excess chromate would begin to oxidize the PDCA, and itself would be reduced to $Cr^{3+}$. The $Cr^{3+}$ could then bind to PDCA that had not yet been oxidized by the hot chromic acid. This would give an artificially high value for $Cr^{3+}$ content. An alternative to this method is to form an anionic EDTA complex with $Cr^{3+}$. Unfortunately, concentrated chromic acid can also partially oxidize EDTA, producing $Cr^{3+}$ in the process, which can give artificially high results. To prevent this from occurring, the pH of the sample should first be raised to about 5.5 using NaOH. EDTA can then be added. Carbonate is known to catalyze the formation of Cr-EDTA.[83] Thus, the sample need not be heated, and the anionic Cr-EDTA can be separated from chromate and be detected by chemically suppressed conductivity.[84]

The AS-7 column has also been used in plating-solution analysis. Originally, it was used for separating polyvalent anions. Some plating baths contain phosphonates as dequest agents. These dequest agents are added to plating baths such as gold baths to bind any calcium and magnesium that may be present from incompletely deionized water. They are eluted with 50 m$M$ nitric acid and are detected by post-column reaction with 1 g/$\ell$ ferric nitrate, followed by UV detection at 330 nm.[85] Thus, a variety of inorganic anions can be determined using an AS-1, 2, 3, 4, or 4A column. Baths containing large, hydrophobic anions can be analyzed using the AS-5.

Organic anions such as aliphatic carboxylic acids have been determined in plating baths using ion chromatography exclusion (ICE). The carboxylic acids are well separated from the inorganic anions with low pKa (such as chloride, nitrate, and sulfate). Anions which have pKa below 2 are almost fully ionized in the 1 m$M$ HCl used as an eluent in ICE. Therefore, they are excluded from the negatively charged pores of the ICE column by Donnan exclusion. Anions with pKa above 3 are only partially ionized, can fit into the pores of the ICE column, and are retained (see discussion in Chapter 2). The higher the pKa, the less ionized the carboxylic acid and the longer it is retained on the ICE column.

Most all the published plating applications for ICE use the HPICE AS-1 or the "standard" ICE separator and the packed-bed suppressor (in the silver form). For example, citrate and succinate were determined in an electroless nickel bath.[76] In electroless copper baths, formaldehyde is used as the reducing agent to cause electrodeposition of Cu. The formaldehyde is converted to formic acid during the copper plating. The age of the bath can be estimated by determining the formate. This can be done using an HPICE AS-1.[86] Carbonate can be determined in a gold cyanide bath using the HPICE AS-1 and a 5-m$M$ HCl eluent as shown in Figure 30. Under these conditions, carbonate elutes in about 14 min (k' about 1.6). Boric acid, which is used as a mild buffer in Watt's nickle, was determined in a Ni/Fe bath using ICE. Because boric acid has a low specific conductance, it must be converted to a higher conductive form. This is done by including 0.1 $M$ mannitol in the eluent. A borate-mannitol complex forms. The pKa of the complex is lower than that of free boric acid. As a result, the mannitol complex is more fully ionized after it comes off the suppressor column, and has higher conductivity. Another additive in Watt's nickle, saccharin, can also be determined by ICE.[75] The HPICE AS-1 has also been used to analyze ethanol-based baths for formate, acetate, and tartarate.[87] The HPICE AS-1 column is quite stable to the ethanol in the samples. It should also be emphasized that fiber suppressors are more available for HPICE, but were not available before these applications were written. It is probably more convenient to perform these analyses using either the anion-fiber suppressor (AFS-II), or the ICE micromembrane suppressor. In addition, a new HPICE column, the AS-5, has been developed since these reports were published. This is a novel type of column which separates weak acids by ion exclusion, dimer formation, and by hydrogen bonding (see Chapter 2 for a more detailed explanation). It has a different selectivity than any of the other HPICE separators. It is quite possible that some of the previously reported methods could be improved by using the HPICE as AS-5. The AS-5 could also be used together with the AS-1 to identify unknown organic acids in plating baths. Since they have different selectivities for organic acids, a sample could be injected on the AS-1 and then on the AS-5. If the retention times of the unknown

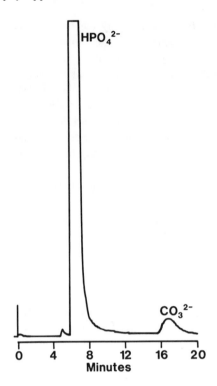

FIGURE 30.    Carbonate in a gold cyanide plating bath. Column: ICE, conductivity detection.

acid on each chromatogram can be matched to a standard on the same columns, the identity of the acid can be confidently confirmed.

Some anions, especially metal-cyanides, are not amenable to ion exchange or ICE. Such large hydrophobic anions are strongly retained on the AS-1, 2, 3, 4, and 5, along with the ICE separators. They are more easily determined by mobile-phase ion chromatography (MPIC). Many gold plating baths use potassium gold cyanide, $KAu(CN)_2$, as the source of gold. Thus, the gold exists as the anionic $Au(CN)_2^-$. This gold-cyanide complex can be separated from dequest agents and other bath components by MPIC. An ion-pair reagent such as tetrapropyl ammonium hydroxide, is put in the eluent along with 10 to 20% acetonitrile. A neutral ion pair forms between the $TPA^+$ and $Au(CN)_2^-$. This ion pair is retained on the MPIC column, and is detected by chemically suppressed conductivity as shown in Figure 31. A packed-bed anion suppressor was used in this. Today, either an AFS-II, or an anion-micromembrane suppressor could be used. This would eliminate the need for continual off-line regeneration of the suppressor. The AFS-I would have been damaged by the acetonitrile and TPAOH in the eluent. The newer AFS-II and micromembrane suppressor are stable to this eluent, and can be used in the analysis. It is also possible to detect trace amounts of $Au(CN)$ (i.e., the cyanide complex with $Au^{2+}$. Increasing amounts of this oxidized $Au^{2+}$ causes plating efficiency to suffer. Even if gold baths do not contain cyanide, they can be diluted into a solution containing cyanide and the gold determined by MPIC. Since one should dilute the bath until the gold concentration falls to about 5 to 20 ppm, little KCN is needed to fully complex the gold, causing limited health hazard.

Cobalt is often added to gold-plating solutions to increase the hardness of the electrodeposit. It is usually added as the cobaltous ($Co^{1+}$) ion. Once oxidized in the presence of cyanide to the very stable $Co(CN)_6^{3-}$ anion is formed. In the oxidized form, cobalt can no longer be plated with the gold, and the quality of the electrodeposit deteriorates.[76] The concentration of $Co(CN)_6^{3-}$ can be determined simultaneously with the $Au(CN)_2^-$ by MPIC

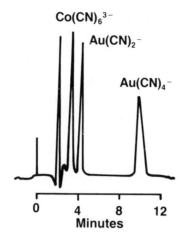

FIGURE 31. Gold I and III complexes in a gold plating bath. Column: MPIC.

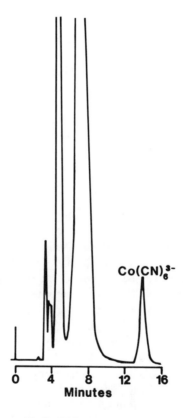

FIGURE 32. Cobalt cyanide $Co(CN)_6^{3-}$ in a phosphate-citrate buffered gold plating bath.

using a 10% acetonitrile/2 $m$M TBAOH (tetrabutyl ammonium hydroxide) eluent as shown in Figure 31. It can also be determined as a trace component in a phosphate/citrate-buffered gold plating bath as shown in Figure 32.

MPIC is also quite useful in determining organic additives to plating solutions. Depending on the type of organic being determined, conductivity or electochemical detection may be useful. Quite often, surface-active agents (detergents) are added to the plating solutions. For anionic detergents, MPIC is required, because most detergents would be strongly retained on anion separators, or ICE separations. Using MPIC, however, linear alkyl and benzene sulfonate detergents can be separated as described previously.[88]

A similar example is the determination of a disulfide anionic brightener in a bright copper sulfate/sulfuric acid bath. UV detection is used because the large sulfate peak would swamp a conductivity detector.[3] The other peaks in the chromatogram were not identified. This is often the case with plating solutions where unknown organics are added to the bath. By monitoring the sizes of the various peaks when doing MPIC analysis, changes are often seen during increased usage of a bath. It is quite useful to correlate these to changes in brightness of the electrodeposit as measured by a Hull cell. If a Hull-cell test indicates that the electrodeposit is not bright enough, carbon treatment could be used to clean the bath of any organic breakdown products. The bath should be analyzed for brightener content (using MPIC in most cases) to determine how much replenishing solution may need to be added. After carbon treatment, the concentrations of the various organics are expected to decrease. Hopefully, harmful breakdown products (i.e., new peaks appearing in the MPIC chromatogram of used bath) are selectively removed by carbon treatment. An MPIC analysis afterward can be used to compare the relative amounts of brightener still present in the plating solution. This information can be used to instruct platers on how much brightener needs to be added back into the bath.

The example of the brightener in the acid copper bath illustrates an important point. The brightener is present at low concentration, so that the bath is only diluted 1/2. The sulfuric acid content can be as high as 2 to 3% in this 1/2 dilution. Such a sample could damage a silica-based column and possibly corrode metal fittings on an HPLC. Thus, even though many compounds that are separated by MPIC could also be separated by reverse-phase HPLC using an octadecyl-silica column, the silica-based column in a metallic HPLC can only be used to analyze noncorrosive samples. It may be possible to neutralize the sulfuric acid in the acid copper bath by adding NaOH, but this would add a time-consuming sample-preparation step and would produce a sample with a very high salt concentration. This high salt concentration might be expected to noticeably alter the chromatographic properties of the organics.

MPIC can also be used to monitor the concentrations of organic cations, i.e., amines. One of the earliest published applications was the separation of mono-, di-, and triethanolamine.[34] They are separated as ion pairs. Hexane sulfonic acid (2 m$M$) is used as the ion-pair reagent. The eluent contains no acetonitrile. The ethanolamines were detected by suppressed conductivity using a packed-bed cation suppression. Today, the cation fiber suppressor of the MPIC micromembrane suppressor can be used. Diethanolamine is determined in electroless copper baths using this method.

Other amines can also be determined by MPIC. Aliphatic amines, because of their very low UV absorptivity, are best detected by chemically suppressed conductivity. Aromatic amines, on the other hand, do have appreciable absorptivity and can be detected with a UV detector. However, some aromatic amines also have appreciable specific conductance and can also be detected by chemically suppressed conductivity. In fact, it is useful to install a conductivity and a UV-visible detector in series to distinguish between aliphatic amines (no UV peak) and aromatic amines (UV peaks and some conductivity peaks) and inorganic cations.

It should be added that post-column reaction with *o*-phthalaldehyde followed by fluorescence detection is also possible for primary amines. Although usually used for amino acids, this detection scheme has appeared at least once for polyamines.[89]

MPIC is also quite useful for determining nonionic organic additives. (When used for this application, no ion-pair reagents are used and conductivity is no longer used for detection. In this mode, MPIC is quite analogous to reversed-phase HPLC, except that the column is packed with polystyrene/divinylbenzene resin. It is also important to realize that the absence of metallic components is also implied in MPIC. The corrosion-resistant nature of the resin and hardware is useful when analyzing highly acidic or alkaline-plating samples.

This is a relatively new area with limited reports published to date. One application is the determination of Triton® X-100 in an alkaline cleaning solution and in a tin/lead plating bath.[90] The Triton® detergent is separated from other bath components using a 70% acetonitrile/30% water isocratic eluent, and a UV detector at 254 nm is used for detection. Triton® X-100 is a member of the general class of nonionic-called alkyl phenol polyethoxylates. This class of detergents includes the Triton® and Igepal® detergents. They differ from each other in that the alkyl group can be either *n*-nonly (C-9) or a highly branched octyl (i.e., 1,1,4,4-tetramethylbutyl) and in the length of the ethoxylate-repeating unit. Triton® X-100 has the highly branched octyl group. Moreover, Triton® X-100 is not a pure compound, but instead is a mixture of oligomers in which there can be anywhere from 4 to 12 ethoxylate units. MPIC is capable of separating these oligomers using a 60% acetonitrile/ 40% water eluent. The same MPIC column can separate Triton® X-100 from Triton® N-101, which also has an average of eight ethoxylates, but has the *n*-nonyl alkyl group. MPIC is also capable of separating Igepal® CA-210 from Igepal® CO-210, which also differ, in that one has the *n*-nonyl and the other has the branched octyl group attached to the phenolic group. The detergents having the *n*-nonyl group are more strongly retained on the polystyrene/ divinylbenzene-based MPIC column than are the detergents having the highly branched octyl group. The detergents having more ethoxylate-repeating units are more polar, and are less strongly retained on the MPIC column. Thus, MPIC can be used to distinguish between different types of nonionic detergents, and is capable of their determination in highly corrosive plating solutions. In most plating baths, the detergent is present at such low concentrations that the sample cannot be diluted very much. Thus, the NaOH of fluoroboric acid in alkaline cleaners or tin/lead alloy plating bath would be so concentrated that it could damage silica-based columns such as the RPIC column. Moreover, the UV detector shows good specificity for the detergent, since the ionic components in these baths do not absorb near 254 nm. Several other nonionic organics have been separated by MPIC. The dihydroxybenzenes, catechol, resorcinol, and hydroquinone were separated using 1:1 methanol/water (v/v).[91] Detection was by UV at 254 nm. It has also been shown that phenols can be separated by MPIC and RPIC columns, and detected by an electrochemical detector using a glassy carbon electrode.[92] It should be noted that each of these dihydroxybenzenes are good ligands for transition metals, and if present in plating solutions, may exist as metal-ligand complexes. It was shown that the copper-dihydroxybenzene complexes were not dissociated during the chromatography and produced separate peaks well separated from free ligand. Thus, if an analyst were to inject known dihydroxybenzene standards and compare their retention times to peaks from a plating-sample chromatogram, the analyst might conclude, incorrectly, that no resorcinol, catechol, or hydroquinone were present. In fact, they could be present as copper complexes. This simply illustrates the point that the presence or absence of an organic (or any other compound) should be tested not simply by running a standard, but by spiking the sample with the compound in question. Methods development for organics using MPIC has been slow because the technique was originally developed for determining large, hydrophobic anions that are strongly retained on anion separator columns. Also, methods based on reversed-phase HPLC using silica-based columns (i.e., octadecyl-silica) already exist for many of the organics (including dihydroxybenzenes) present in plating solutions. However, the highly corrosive nature of the plating solutions could cause problems with the silica-based columns and the metallic fittings of the HPLC.

Methods development for nonionic, UV-absorbing organics in plating solutions is quite easy. The polystyrene/divinylbenzene resin and the nonmetallic fittings on the ion chromatograph are quite stable to daily use of up to 90% acetonitrile. In practice, however, very few compounds that are soluble in the highly ionic plating baths require over 70% acetonitrile for their effective elution off the MPIC column. Because the major components of most plating baths are ionic with little or no UV absorbance, the UV detector is quite specific for the organics.

One other application to plating bath analysis is mono- and divalent cation determination. Two separate CS-1 columns were originally required to determine mono- and divalents. The very strong *m*-phenylenediamine eluent needed for barium, calcium, magnesium, and strontium cannot be conveniently washed off the CS-1 to permit separation of lithium, sodium, ammonium, and potassium. Presently, however, the CS-3 column and the cation micromembrane suppressor can be used for simultaneous separations of mono- and divalents in one run. Instead of dedicating one CS-1 to monovalents and another CS-1 to divalents, only one CS-3 is needed.

Plating solution analysis can be automated using either of two schemes. In the first, the analyst must still sample the bath, dilute it, and load it into the autosampler. In the other scheme, a Series 8000® ion chromatograph can automatically sample the bath, dilute it, and perform an analysis. The Series 8000® can accept up to 12 sample streams and perform two different types of analysis, i.e., anions and cations. All the plating chemist need do is periodically replenish the regenerant solutions.

In the first scheme, the analyst must still sample the bath, dilute it, and load the samples into the autosampler. A Series 2000® or 4000® ion chromatograph would be used in this scheme, and would provide far more diversity than the Series 8000®. Up to six different separator columns can be installed into a dual chromatography module in the Series 2000® or Series 4000®. Electrochemical, conductivity, and UV-visible detectors can be used. The analyst need only mix-and-match the appropriate separator and detector for the specific samples being analyzed. For example, an AS-4A and AS-5 column can be installed on System 1 Valve A. Programs can be written on the microprocessor or computer to automatically select the AS-4A for most inorganic-anion analyses, and the AS-5 for analyses involving large, hydrophobic anions, or requiring gradient elution. The effluents off these columns are both directed to the same B valve which controls regenerant flow. Only one anion micromembrane (or any other) suppressor and one conductivity detector need be used. The third column on System 1 could be an MPIC column that is dedicated to determining organic additives. This column would be selected by manually by-passing the A valve in a 1-min operation. The effluent off the MPIC column could be directed to an MPIC micromembrane suppressor, a conductivity cell, and a UV-visible detector connected in tandem. If desired, a post-column reactor could also be installed. On System 2, a CS-3 and an HPICE AS-5 column could both be connected to the A valve. Again, the microprocessor or computer can select the appropriate column automatically. Both columns use acidic eluents, but it would probably be best to first wash the System with the 1 m*M* HCl eluent used for ICE separations, when finished with CS-3 separations. Separate suppressor columns would be required for cation (CS-3) and organic acids (ICE). Only one conductivity cell is required, however. One possible arrangement would be to connect the cation suppressor to the B valve, and use a packed-bed suppressor for ICE. The ICE effluent can also be directed to the B valve so that no matter which analysis is being performed, the effluent off the B valve is directed to the same conductivity cell (3). The third column on System 2 could be a CS-5. Thus, one dual-system ion chromatograph could be used to analyze plating baths for anions, cations, organics, and metals in an automated mode. All the analyst would need to do is dilute the samples and load them into an autosampler. After recalling previously written programs, fully unattended analysis begins. The concentrations of plating-bath components are calculated from calibration data and printed in report format.

## IV. OTHER APPLICATIONS

Of course, there are many other applications besides those in the life sciences, environmental protection, and electroplating. These have been summarized in a number of books and reviews since ion chromatography was first introduced.[26,27,77,89] As mentioned several times before, the columns and detectors that are described in each subsequent review are constantly updated. Two reviews published in 1984 present an overview of industrial applications[93,94] Most of the diverse capabilities of ion chromatography were reviewed briefly, but in good detail in 1985.[77,95,96] None of these, however, described the use of gradient elution, which will probably find a number of interesting applications in the near future. Industrial applications of ion chromatography include (1) water analysis in the manufacture of chemicals, such as boiler-water composition (the determination of phosphate to control scale formation and the determination of sulfite, which is an important oxygen scavenger and anti-oxidant), (2) the analysis of organic dyestuffs (phosphate is often added as a stabilizer), (3) control of the manufacture of aniline (to monitor the ratio of nitrate to nitrite), and (4) the monitoring of corrosion inhibitor (nitrite) in machine cutting fluids.[93] In a separate review, the following applications were discussed (1) the determination of chloride and sulfate in the steam-water cycle of an ammonium plant, (2) the determination of trace levels of the highly corrosive chloride in the steam-water cycle of electric power plants, and (3) the monitoring of mining-process streams and effluents for precious minerals.[94] Other areas in which ion chromatography has provided useful information include the pulp and paper industry, beer and wine analysis, analysis of chemicals for purity, elemental analysis in organics and polymers, analysis of detergents, and the manufacture of printed circuits used in computers.

An example of an application that is important to a variety of industries is the monitoring of water that manufacturing facilities discharge into the public water. As shown in Figure 33, ion chromatography can be used to monitor the concentrations of metals in plant secondary waters. In some cases, it is important to differentiate the oxidation state of the metal. For example, chromium is quite toxic and is carcinogenic when present in the +6 oxidation state (i.e., dichromate or chromate). This can be done by ion chromatography. As discussed in Chapter 3, most transition metals are detected by converting them to strong chromophores. This is accomplished by a post-column reaction with pyridyl azo resorcinol (PAR). When determining Cr(III), however, the post-column reagent diphenylcarbazide is used. In addition, pyridine dicarboxylic acid (PDCA) is added to the eluent to accelerate the kinetics of binding of Cr(III) to the reagent.

When reading applications in the ion-chromatographic literature, care should be taken to not always assume that the presence of a peak eluting at the same time as chloride, is necessarily due to nitrate. As demonstrated by Littlejohn and Chang,[96] ion-chromatography columns can also be used to separate nitrogen and phosphorous-containing anions. They demonstrated the separation of the nitrogen sulfonates hydroxyimidosulfate (HIDS), hydroxysulfamate (HSA), nitridotrisulfate (NTS), imidodisulfate (IDS), and N-nitrosohydroxylamine-N-sulfonate (NHAS). IDS, HIDS, and NHAS were separated on an AS-4 separator column using 18 mM carbonate eluent. Such a strong eluent is required to elute the strongly retained NHSA. Sulfamate and HSA, on the other hand, are weakly retained, and require 1.5 mM bicarbonate to be separated. These compounds are important because they appear in a flue-gas scrubbing liquor.[96]

One of the more interesting applications for ion chromatography is in the analysis of beer.[97] The first step in making beer is when barley, rice, and other grains are soaked in warm water. This process is called mashing, the resulting solution is called sweet wort. The sweet wort is then treated with hops to begin the hydrolysis of the large molecular weight carbohydrates present in the original grains. The resulting solution is called hopped wort.

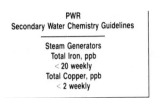

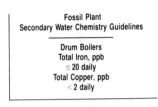

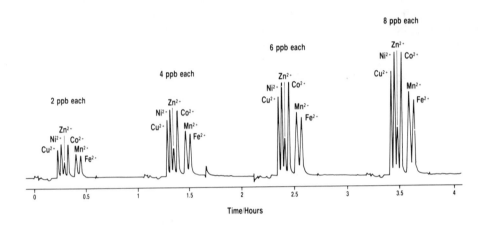

FIGURE 33. Trending data: on-line ion chromatographic system for monitoring corrosion products in plant secondary waters.

Yeast is then added to ferment the smaller carbohydrates present in the hopped wort. One of the keys to brewing a good beer is to control these various steps. Of central importance is the concentration of carbohydrates. In the starting material (grains), the raw material is large carbohydrates. The number of simple sugars that are chemically linked in glycosidic bonds is called the degree of polymerization of the carbohydrate. The originally large carbohydrates are thus called polysaccharides. These polysaccharides are hydrolyzed to lower molecular weight oligosaccharides in the initial worts. To convert the sometimes bitter-tasting oligosaccharides to alcohol and sweeter mono-, and disaccharides, further hydrolysis and fermentation is required. To ensure proper beer production and quality taste, the concentrations of the simple sugars and oligosaccharides in the different steps at the brewery, are necessary.

To begin with, it is quite easy to dilute, filter, and inject beer to determine the anions present. This has been done since the earliest days as shown in Figure 34. Although the AS-1 column used in this separation is adequate, the same analysis can be performed much faster, and with better resolution, using columns such as the AS-4A. Since the introduction of the pulsed amperometric detector, beer has been analyzed for oligosaccharides,[97] as shown if Figure 35. It should be noted that the carbohydrates are being separated as anions. This may seem curious, since chromatographers, organic chemists, and biochemists usually do not consider carbohydrates to be ionic. They have extememly high pKa values (12 to 14), but in the presence of $0.15\ M$ NaOH, however, they are ionized. This is too strong an eluent for most separator columns. The AS-1, 2, 3, 4, 4A, and 5 columns all have too few ion-exchange sites (i.e., low capacity) to enable the use of eluents as concentrated as the $0.15\ M$ NaOH plus $0.15\ M$ sodium acetate used in Figure 35. The AS-6 column, however, has a higher ion-exchange capacity and is not overloaded by this eluent.

To separate the lower molecular weight carbohydrates in beer, a slightly weaker eluent

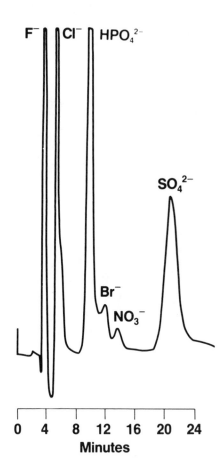

FIGURE 34.    Anion determination in beer. Sample preparation: dilute, filter, and inject.

is used, 0.15 $M$ NaOH, as shown in Figure 36. It is important to control the relative amounts of these mono- and disaccharides to obtain proper fermentation and taste. Alcohols can also be determined in beer.[97] They can be separated by ion chromatography exclusion (ICE), and detected by pulsed amperometry. Organic acids in beer can also be separated by ICE, but they are detected by chemically suppressed conductivity. In the review in 1983, separate cation-separator columns were used to determine mono- and divalent cations.[97] Today, this same separation requires only one column, the CS-3, and the modern cation-micromembrane suppressor. An example of the advancing technology in ion chromatography can be seen in this review[97] however, because the same anion analysis shown in Figure 34 was performed in 7 min using the more modern AS-4 column.

Another application for ion chromatography is in the analysis of highly ionic solutions for trace ionic contaminants. This is especially noteworthy because the preponderance of papers which have appeared on ion chromatography describe low detection limits in very dilute solutions. When analyzing highly ionic solutions, such as brines, acid pickling baths, or even 0.5% NaOH, this same sensitivity should be remembered. It is easy for the analyst to become accustomed to samples that require only analysis for the major ionic species. Such samples need only be diluted to the milligram per liter concentration range, and can still be detected using a 3- or 10-$\mu$S scale. The baseline is almost always quite noiseless using these scales. The analyst should avoid the tendency to use these higher scales when analyzing highly ionic solutions for trace ions. The separator columns can become overloaded by the primary ions if the sample is not diluted enough. The need to either dilute the sample or remove interfering major ionic components has been discussed by several workers.[79,98-]

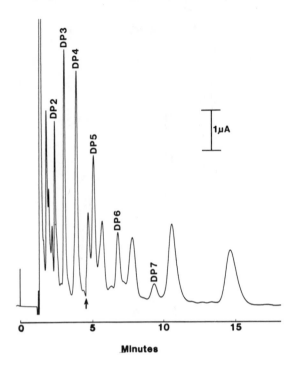

FIGURE 35.    Oligosaccharides in beer. Pulsed amperometric detection.

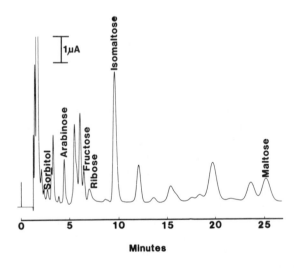

FIGURE 36.    Mono and disaccharides in beer (0.15 *M* NaOH eluent). Pulsed amperometric detection.

101 Dulski[98] analyzed acid pickling baths that are used to prevent oxide scale on stainless steel and specialty alloys. The pickling baths dissolve metals which can form metal-fluoride complexes. Ion chromatography can be used to determine the metal and free fluoride concentrations. Cox and Tanaka[99] used a cation exchange resin in the proton form to pretreat NaOH and sodium carbonate samples before ion chromatographic analysis.[99] This converted the hydroxide and carbonate to water and carbonic acid, respectively. This work was done before the introduction of micromembrane suppressors, however. The fiber suppressor used by Cox and Tanaka has limited ion exchange capacity, and can only convert a small amount

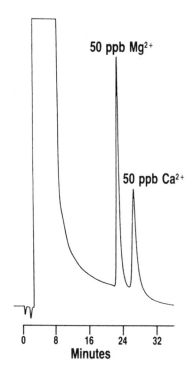

FIGURE 37.    Cation standards in 10% brine.

of the hydroxide or carbonate in the sample to their low conductive forms (water and carbonic acid). The anion-micromembrane suppressor, however, has a much larger dynamic suppression capacity, and can be used in lieu of sample pretreatment to analyze samples as concentrated as 0.5% NaOH without dilution.[102] Another chemical which must be analyzed for purity is boric acid, or boric oxides. Boric acid can be determined in dilute solutions by taking advantage of the fact that it forms a highly conductive complex with mannitol. The boric acid is separated on an ICE column using a 1-m$M$ HCl/0.1 $M$ mannitol eluent and chemically suppressed conductivity detection.[101,103] To determine high levels of boric acid, water can be used as the eluent on an ICE separator, but sensitivity is low.[103] Since the advent of the micromembrane suppressor, trace anions in boric acid can be determined accurately[104] without any complicated matrix effects that were alluded to before the micromembrane suppressor was available.[100]

Another highly ionic sample requiring analysis for trace ionic components is brine. The manufacturers of high-purity chemicals by the chlor-alkali membrane process require high-purity brines. Metals, especially calcium and magnesium, must be carefully controlled. Although other detection methods exist, chemically suppressed conductivity is easiest for calcium and magnesium.[105] To perform trace analysis, however, it may seem necessary to preconcentrate the metals present. An alternative is to isolate the metals by complexing with iminodiacetate. The iminoacetate is especially effective when bound to a solid resin substrate. With this preconcentration and subsequent matrix elimination with the resin, nanogram per liter (parts per billion) levels of calcium and magnesium can be separated and detected as shown in Figure 37. This separation was accomplished on a CS-3 column using a step gradient elution.

Another example of sample pretreatment in analyzing highly ionic solutions was described by Chriswell et al.[106] They described the determination of total sulfur in spent caustic by oxidation with hydrogen peroxide. Caution is required, however, in the addition of hydrogen peroxide to highly basic solutions. The results for total sulfur were reproducible and accurate, based on calculations of total sulfur balance.

Ion chromatography is also quite useful in the other extreme, i.e., in analyzing highly purified, deionized water. Ion chromatography can be used with a mini-leak detector to monitor cooling water in power plants to protect expensive, large turbine blades in condensers.[107] This is important because of the high water pressures used. For example, if there is only 8 μg NaCl per kilogram of water, it can be problematic. With 2000 tons of water being used per hour, this would give 384 g NaCl per day.[107] A detection limit of 0.1 μg/kg water was reported.

High-purity water is also important in manufacturing integrated circuits. If chloride is present in the circuits in a hand-held calculator, or especially in a main frame computer, serious and expensive damage can result. Chloride can migrate and cause short circuits. Thus, workers at IBM have reported the analysis of water at the primary central production station and in the secondary polishing station in their integrated circuit manufacturing facility.[108]

Another analysis that is important in the chemical industry is the determination of silica. It can be determined by converting it to the anionic hexafluorosilicate.[109] Borates which are used in glasses, fiberglass insulation, pharmaceuticals, household-cleaning products, agriculture, and the nuclear power industry must be analyzed for silica. The conversion of silica to hexafluorosilicate is accomplished by reaction with HF. The hexafluorosilicate was separated on an AS-1 column using 3 mM sodium bicarbonate eluent,[109] but certainly the AS-3 or even AS-4 or 4A columns can be used with an HPIC system.

Ion chromatography is also useful in analyzing organics and polymers. It has been used to determine the acetyl and propionyl end groups in poly-epsilon-caprolactam. The end groups are hydrolyzed to acetate and formate which can be separated on an ICE column.[110] Ion chromatography is especially useful in elemental analysis of organics and polymers. The covalently bound halogens, sulfur, and phosphorous can be pyrohydrolyzed in a Parr bomb or a Schoniger combustion flask[111] to their anionic forms (halides, phosphate, and sulfate). These anions can then be separated and detected by ion chromatography. Some materials do not burn well even under the oxygen pressures used in a Parr bomb. In such cases, sodium fusion may be necessary to convert the covalently bound halogens, phosphorous, and sulfur to their anionic forms.[112]

Another class of organics that can be analyzed by ion chromatography is color additives to foods, drugs, and cosmetics.[113,114] In one of the earliest applications, azo-sulfonated organic dyes were analyzed for anion content. Chloride, sulfate, and occasionally fluoride or phosphate were determined. To determine iodide in these early studies, a short separator column (called a precolumn at the time) was used.[113] Iodide is strongly retained on a full-sized separator column, but can be eluted off a short precolumn if a strong eluent is used. These studies were followed by more detailed analyses. Iodate, glycolate, carbonate, pyruvate, formate, acetate, chloroacetate, and bromate were all found to elute between fluoride and chloride. To effect separation, a weaker eluent, such as 1.5 mM sodium bicarbonate is required. Iodate, formate, acetate, chloride, phosphate, bromide, nitrate, sulfate, and iodide were all found in color additives. This report[114] is very useful for one other thing. It contains one of the most extensive listings of anions and their relative retention times off the original AS-1 column. Most probably, the elution orders are not very different on the AS-3 or 4 columns.

Another area for ion-chromatography applications is detergents. One of the first applications for the original (single voltage) electrochemical detector was the determination of hypochlorite in bleach. The active ingredient in bleach is the hypochlorite, but chloride and chlorate are also present. Phosphonates have also been determined by ion chromatography. They are strongly retained on most anion separators (such as the AS-1, 2, 3, 4, 4A, and 5). To be eluted, they require a strong eluent (such as 0.15M nitric acid). As seen before, the AS-6 column has the highest anion-exchange capacity and is best suited for such concentrated eluents. Thus, polyphosphonates can be separated on an AS-6 column. This was first done before the advent of micromembrane suppressors and gradient elution. As a result, no suppressor was used, but instead, post-column reaction with ferric nitrate followed by detection at 410 nm was used. For more details on the use of this post-column reaction to detect polyvalent anions, see Chapter 3. Detergents are not always inorganic. More often, they are organic. Commercial shampoos, for example, may contain alkyl sulfates such as sodecyl sulfate and tetradecylsulfate. These and other similar alkyl sulfates would be strongly retained if injected on an anion separator column, and especially so on an ICE column. They can be eluted off an MPIC column, though. In fact, nonionic detergents such as Triton® X-100 can be eluted off an MPIC column using 70% acetonitrile.[86]

Another area for ion chromatography is the analysis of geological samples. Wilson and Gent,[115] used sodium carbonate fusion to extract chloride from silicate rocks. They used an AS-3 and bicarbonate eluent to determine the chloride content. This analysis was useful in evaluating geothermal wells, oil shales, and soil extracts. Nadkarni and Pond[116] used oxygen combustion followed by ion chromatography of the pyrohydrolysate to determine sulfur in coal and oil shale. They compared ion chromatographic results with those obtained by X-ray fluorescence, polarography, and an ion-selective electrode, to verify accuracy.

One other application for ion chromatography is the pulp and paper industry. The functional areas of this industry include pulping, bleaching, paper making, corrosion engineering, steam and energy production, pollution control, and industrial hygiene. Anions determined include sulfate, sulfide, thiosulfate, sulfite, phosphate, chloride, bromide, fluoride, carbonate, oxalate, glycolate, acetate, formate, lactate, calcium, magnesium, barium, strontium, hydrazine, hypochlorite, phenols, nitrite, nitrate, cyanide, chlorite, chlorate, and perchlorate.[117-120] The first step is the preparation of the Kraft smelt, which is a solid. Sample preparation included simply soaking in deionized water, and then injection. The next sample analyzed is the Kraft black liquor. The relative amounts of these sulfur species must be carefully controlled. To determine sulfide, sulfate, sulfite, and thiosulfate can require many hours of applied labor if they are determined individually by wet-chemical methods. Using ion chromatography, these anions can be separated and determined simultaneously in one run as shown in Figure 38. Both chemically suppressed conductivity and electrochemical detection were used. Many of these analyses can be performed using an automated ion chromatograph. Instead of the approximately 24 hr of applied labor and 3 to 5-day period required for wet-chemical analysis,[117] only a few minutes of applied labor and 3 to 5 hr of instrument time would be required to analyze ten such samples with an automated ion chromatograph. Aliphatic organics such as lactate, formate, and acetate can be determined in Kraft black liquor using ICE. The same sulfur speciation shown in Figure 38 can also provide quite useful information about a Kraft smelt. In addition to the manufacturing processes, the products from a bleach plant can be monitored for hypochlorite, chloride, and chlorate using simultaneous conductivity and electrochemical detection.

Thus, ion chromatography has quite a number of applications, only some of which have been discussed here. The separator columns have been used for years in some laboratories, with impressive reliability. All that is required to maintain column lifetime is to use some common sense. Dilute samples as much as possible, filter all samples, and use a guard column. If these precautions are observed, automated ion chromatography can be a reliable, cost-effective analytical method. It has proven to be useful in helping to solve important industrial, social, and scientific problems.

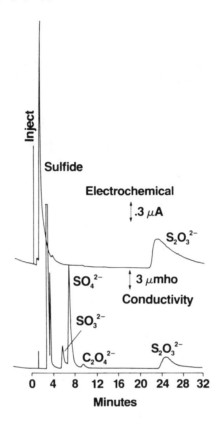

FIGURE 38.    Kraft black liquor. Simultaneous conductivity and electrochemical detection.

# REFERENCES

1. **Henderson, S. K. and Henderson, D. E.,** Enhanced UV detection of sugar phosphates by addition of a metal complex to the HPLC mobile phase, *J. Chromatogr. Sci.,* 23, 222, 1985.
2. **Henderson, S. K. and Henderson, D. E.,** Reversed-phase ion pair HPLC analysis of sugar phosphates, *J. Chromatogr. Sci.,* 24, 198, 1986.
3. **Smith, R. E. and Smith, C. H.,** Automated ion chromatography, applications to plating analysis, *Liq. Chromatogr. Mag.,* 3, 578, 1985.
4. **Mirna, A., Wagner, H., Klotzer, E., and Fausel, E.,** Zur Anwendung der Ionenchromatographie bei der Untersuchung von Fleisch und Fleischwaren. (Applications of ion chromatography in the study of meat and meat products), *Lebensmittelchem. Gerichtl. Chem.,* 38, 18, 1984.
5. **Sawicki, E.,** Application of ion chromatography to the analysis of environmental genotoxicants, in *Ion Chromatographic Analysis of Environmental Pollutants,* Ann Arbor Science, Ann Arbor, Mich., 1979.
6. **Rehfeld, S. R., Loken, H. F., Nordmeyer, F. R., and Lamb, J. D.,** Improved ion chromatographic method for determining magnesium and calcium in serum, *Clin. Chem.,* 26, 1232, 1980.
7. **Curry, S. H.,** Drug assay in therapeutic monitoring, *Trends Anal. Chem.,* 5, 102, 1986.
8. **Cole, D. and Scriver, C.,** Microassay of inorganic sulfate in biological fluids by controlled flow anion chromatography, *J. Chromatogr. Biomed. Appl.,* 225, 359, 1981.
9. **Mahle, C. J. and Menon, M.,** Determination of urinary oxalate by ion chromatography — preliminary observation, *J. Urol.,* 127, 159, 1982.
10. **Robertson, W. G., Scurr, D. S., Smith, A., and Orwell, R.,** Prevention of ascorbic acid interference in the measurement of oxalic acid in urine by ion chromatography, *Clin. Chim. Acta,* 140, 97, 1984.
11. **Zerwekh, J., Drake, E., Gregory, J., Griffith, D., Hofman, A., and Menon, M.,** Assay of urinary oxalate: six methodologies compared, *Clin. Chem.,* 29, 1977, 1983.

12. Dionex Corporation, Determination of oxalate in urine, Application Note 36, 1982.
13. **Ames, B.,** Dietary carcinogens and anti-carcinogens, *Science,* 221, 1256, 1983.
14. **Rich, W., Johnson, E., Lois, L., Kabra, P., Stafford, B., and Marton, L.,** Determination of organic acids in biological fluids by ion chromatography: plasma lactate and pyruvate and urinary vanillylmandelic acid, *Clin. Chem.,* 26, 1492, 1980.
15. **Pisano, J. J., Richard, J., and Abraham, D.,** Determination of 3-methoxy-4-hydroxymandelic acid in urine, *Clin. Chim. Acta,* 7, 285, 1962.
16. **Lebel, A. and Yen, T. F.,** Ion chromatography for the determination of metabolic patterns of sulfate-reducing bacteria, *Anal. Chem.,* 56, 807, 1984.
17. **Busman, L. M., Dick, R. P., and Tabatabai, M. A.,** Determination of total sulfur and chlorine in plant materials, *Soil. Sci. Soc. Am. J.,* 47, 1167, 1983.
18. **Basts, N. T. and Tabatabai, M. A.,** Determination of total potassium, sodium, calcium, and magnesium in plant materials by ion chromatography, 49, 76, 1985.
19. **Oikawa, K., Saito, H., Sakazume, S., and Fujii, M.,** Behavior of potassium in bread by ion chromatography, *Chemosphere,* 11, 953, 1982.
20. **Menconi, D. E.,** Determination of urea in fertilizers using ion chromatography coupled with conductometric detection, *Anal. Chem.,* 57, 1790, 1985.
21. **Buechele, R. C. and Reutter, D. J.,** Determination of ethylenediamine in aqueous solutions by ion chromatography, *Anal. Chem.,* 54, 2114, 1982.
22. **Edwards, P.,** Ion chromatography — a valuable analytical tool for the food chemist, *Food Technol.,* June, 53, 1983.
23. **Eubanks, D.,** Applications of ion chromatography in agronomy, Pittsburgh Conf., 1986.
24. Dionex Corporation, Analysis of organic and inorganic ions in milk products, Application Note #10, 1980.
25. **Anderson, C., Warner, C. R., Daniels, D. H., and Padgett, K. L.,** Ion chromatographic determination of sulfites in foods, *J. Assoc. Off. Anal. Chem.,* 69, 14, 1986.
26. **Sawicki, E., Mulik, J. D., and Wittgenstein, E.,** *Ion Chromatographic Analysis of Environmental Pollutants,* Ann Arbor Science, Ann Arbor, Mich., 1977.
27. **Mulik, J. D. and Sawicki, E.,** Ion Chromatographic Analysis of Environmental Pollutants, *Vol. 2, Ann*
28. **Sawicki, E.,** Application of ion chromatography to the analysis of environmental genotoxicants, in *Ion Chromatographic Analysis of Environmental Pollutants,* Vol. 2, Mulik, J. D. and Sawicki, E., Eds., Ann Arbor Science, Ann Arbor, Mich., 1979, 1.
29. **Smee, B., Hall, G., and Koop, D.,** Analysis of fluoride, chloride, nitrate, and sulfate in natural waters using ion chromatography, *J. Geochem. Explor.,* 10, 245, 1978.
30. **Steiber, R. and Merrill, R.,** Application of ion chromatography to the analysis of source assessment samples, in *Ion Chromatographic Analysis of Environmental Pollutants,* Vol. 2, Mulik, J. D., and Sawicki, E., Eds., Ann Arbor Science, Ann Arbor, Mich,. 1979, 99.
31. **Slanin, J., Lingerak, W., Ordelman, J., Borst, P., and Bakker, F.,** Automation of ion chromatography and adaptation for rainwater analysis, in *Ion Chromatographic Analysis of Evironmental Pollutants,* Vol. 2, Mulik, J. D. and Sawicki, E., Eds., Ann Arbor Science, Ann Arbor, Mich., 1979, 305.
32. **Liljestran, H. and Morgan, J.,** Chemical composition of acid precipitation in Pasadena, California, *Environ. Sci. Technol.,* 12, 1271, 1978.
33. **Tyree, S., Stouffer, J., and Bollinger, M.,** Ion chromatographic analysis of simulated rainwater, in *Ion Chromatographic Analysis of Environmental Pollutants,* Vol. 2, Mulik, J. D. and Sawicki, J. D., Eds., Ann Arbor Science, Ann Arbor, Mich., 1979, 295.
34. **Galvin, P. and Cline, J.,** Measurement of anions in the snow cover of the Adirondack mountains, *Atmos. Environ.,* 12, 1163, 1978.
35. **Galvin, P., Samson, P., Coffey, P., and Romano, D.,** Transport of sulfate in New York State, *Environ. Sci. Technol.,* 12, 1978; Kasiske, D. and Sonnenborn, M., Analysis of anionic constituents in natural waters by ion chromatography, *Labor Praxis,* 4, 76, 1980.
36. **Itoh, H. and Shinbori, Y.,** Determinations of anions in sea water by ion chromatography, *Bunseki Kagaku,* 29, 239, 1980.
37. **Darimont, T., Schulze, G., and Sonnenborn, M.,** Nitratbestimmung in Trinkwasser mit Hilfe der Ionenechromatographie (Determination of Nitrate in drinking water using ion chromatography), *Fres. Z. Anal. Chem.,* 314, 383, 1983.
38. **Kempf, T. and Sonnenborn, M.,** Chemische Zusamensetzung von Trinkwasser in verschiedenen Gebieten der Bundesrepublik Deutschland (Chemical Composition of drinking water in various regions of the Federal Republic of Germany, *vom Wasser,* 57, 83, 1981.
39. **LeGrand, M., deAngelis, M., and Delmas, R. J.,** Ion chromatographic determination of common ions at ultratrace levels in Antarctic snow and ice, *Anal. Chim. Acta,* 156, 181, 1984.
40. **Rothert, J.,** Use of ion chromatography for analysis of MAPS precipitation samples, *Nat. Tech. Info. Cent. (NTIC) Pub. PNL-SA-8425,* 1980.

41. **Eaton, A. E., Carter, M., Fitchett, A., Oppenheimer, J., Bollinger, M., and Sepikas, J.,** Comparability of ion chromatography and conventional methods for drinking water analysis, *Water Qual. Technol. Conf.,* 1984.

42. **Fitchett, A. W.,** Analysis of rain by ion chromatography, in *Sampling and Testing of Rain,* ASTM STP 823, Campbell, S. A., Ed., *Am. Soc. Test. Mater.,* 1983, 29.

43. **Pyen, G. S., Brown, M. R., and Erdman, D. E.,** Automated ion chromatographic determination of anions in precipitation samples, *Am. Lab.,* May 22, 1986.

43. **Lindberg, S. E., Lovett, G. M., Richter, D. D., and Johnson, D. W.,** Atmospheric deposition and canopy interactions of major ions in a forest, *Science,* 231, 141, 1986.

44. **Johnson, H. A. and Siccama, T. G.,** Acid deposition and forest decline, *Environ. Sci. Technol.,* 17, 294A, 1983.

45. **Eaton, A., Carter, M., Fitchett, A., Oppenheimer, J., Bollinger, M., and Sepikas, J.,** Comparability of ion chromatography and conventional methods for drinking water analysis, *Proc. Water Qual. Tech. Conf.,* 1984.

46. **Mosko, J. A.,** Automated determination of inorganic anions in water by ion chromatography, *Anal. Chem.,* 56, 629, 1984.

47. **Pyen, G. S. and Erdman, D. E.,** Automated determination of bromide in water by ion chromatography, *Anal. Chim. Acta,* 149, 355, 1983.

48. **Ricci, G. R., Shepard, L. S., Colovos, G., and Hester, N. E.,** Ion chromatography with atomic absorption spectrometric detection for determination of organic and inorganic arsenic species, *Anal. Chem.,* 53, 610, 1981.

49. **Vijan, P. N. and Wood, G. R.,** Semiautomated determination of lead by hydride generation and atomic absorption spectrophotometry, *Analyst,* 101, 96, 1976.

50. BWR Owners Group, Water Chemistry Guidelines Committee. BWR water chemistry guidelines — final report. *Elec. Power Res. Inst.,* April, 1984.

51. **Blair, C., Mullinex, J., and Barker, J.,** Water chemistry studies at the Sequoyah nuclear plant using an on-line ion chromatographic analyzer. Paper at the 46th *Ann. Meet. Int. Water Conf.,* 1985.

52. **Potts, M. E. and Potas, T. A.,** Determination of anions in cooling-tower wastewater from coal gasification by ion chromatography, *J. Chromatogr. Sci.,* 23, 411, 1985.

53. **Joyce, R. J., Eubanks, D., and Schibler, J.,** Selective and sensitive determination of total cyanide in waste materials by ion chromatography, *Pittsburgh Conf.,* 1986.

54. **Wilson, D. L., Durham, H. B., and Thurnau, R. C.,** Determination of cyanide in aluminum industrial wastewater by ion chromatographic and spectrophotometric techniques, *LC-GC Mag.,* 4, 578, 1986.

55. **Munger, J. W., Tiller, C., and Hoffmann, M. R.,** Identification of hydroxymethanesulfonate in fog water, *Science,* 231, 247, 1986.

56. **Boyce, D. and Hoffmann, D. R.,** Kinetics and mechanism of the formation of hydroxymethanesulfonic acid at low pH, *J. Phys. Chem.,* 88, 4740, 1984.

57. **Richards, L. W., Anderson, J. A., Blumenthal, D. L., McDonald, J. A., Kok, G. L., and Lazrus, A. L.,** Hydrogen peroxide and sulfur (IV) in Los Angeles cloud water, *Atmos. Environ.,* 17, 911, 1983.

58. **McArdle, J. A. and Hoffmann, D.R.,** Kinetics and mechanism of the oxidation of aquated sulfur dioxide by hydrogen peroxide at low pH, *J. Phys. Chem.* 87, 5425, 1983.

59. **Hoigne, J., Bader, H., Haag, W. R., and Staehelin, J.,** Rate constants of reactions of ozone with organic compounds in water. III. Inorganic compounds and radicals. *J. Water Res.,* 19, 993, 1985.

60. **Warnek, P., Klippel, W., and Moortgat, G. K.,** Formaldehyde in the clean tropospheric air, *Ber. Bunsenges. Phy. Chem.,* 82, 1136, 1978.

61. **Klippel, W. and Warnek, P.,** The formaldehyde content of atmospheric aerosols, *Atmos. Environ.,* 14, 809, 1980.

62. **Mulik, J. D., Todd, G., Estes, E., Puckett, R., Sawicki, E., and Williams, D.,** Ion chromatographic determination of atmospheric sulfur dioxide, in *Ion Chromatographic Analysis of Environmental Pollutants,* Vol. 1, Ann Arbor Science, Ann Arbor, Mich., 1978, 23.

63. **Mason, D. W. and Miller, H. C.,** Measurement of ambient aerosol with analysis by ion chromatography, in *Ion Chromatographic Analysis of Environmental Pollutants,* Vol. 2, Ann Arbor Science, Ann Arbor, Mich., 1979, 193.

64. **Smith, D. L. and Kim, W. S.,** Determination of sulfur dioxide by absorption on solid sorbent followed by ion chromatography, *Am. Ind. Hyg. Assoc. J.,* 41, 485, 1980.

65. **Fuchs, G. R., Lisson, E., Schwarz, B., and Bachmann, K.,** Analyse von Anionen in aerosolen der Luft, (Anion analysis in air aerosols.), *Fresn. Z. Analyt. Chem.,* 320, 498, 1985.

66. **Viswanadham, P., Smick, D. R., Pisney, J. J., and Dilworth, W. F.,** Comparison of ion chromatography and titrimetry for determination of sulfur in fuel oils, *Anal. Chem.,* 54, 2431, 1982.

67. **Dolzine, T. W., Esposito, G. G., and Rinehart, D. S.,** Determination of hydrogen cyanide in air by ion chromatography, *Anal. Chem.,* 54, 307, 1982.

68. U.S. EPA, Methods for chemical analysis of water and wastes, *Environmental Monitoring and Support Laboratory,* Cincinnatti, Ohio, 1979.

69. **Fortune, C. R. and Dellinger, B.,** Sampling and ion chromatographic determination of formaldehyde and acetaldehyde, *Anal. Chem.,* 53, 1302, 1981.

70. **Bouyocos, S. A., Melcher, R. G., and Vaccaro, J. R.,** Collection and determination of sulfuryl fluoride in air by ion chromatography, *Am. Ind. Hyg. Assoc. J.,* 44, 57, 1983.

71. **Wu, W. S., Arai, D. K., Nazar, M. A., and Leong, D. K.,** Determination of nitrites in metal cutting fluids, *Am. Ind. Hyg. Assoc. J.,* 43, 942, 1982.

72. Dionex Corporation, Determination of ethanolamines in refinery water, *Appl. Note* 39, 1982.

73. **Haak, K. K. and Franklin, G. O.,** Rapid analysis and troubleshooting of gold, copper, and nickel plating bath chemistry by ion chromatography, *Am. Electr. Soc. Conf. Proc.,* I-6, 1983.

74. **Haak, K. K.,** The use of ion chromatography in the electroplating industry, *Product Finish.,* 37, 3, 13, 1984.

75. **Smith, R. E. and Smith, C. H.,** Automated ion chromatography. Application to the analysis of plating solutions, *LC Mag.,* 3, 578, 1985.

76. **Haak, K. K.,** Ion chromatography in the electroplating industry, *Plating Surf. Finish.,* 34, Sept., 1983.

77. **Witcofsky, J. R.,** The use of ion chromatography in the monitoring of plating bath constituents, *FACCS Conf. Paper,* 1984.

78. **Pohl, C., Haak, K. K., and Fitchett, A.,** Determination of metal EDTA complexes via ion chromatography, *Pittsburgh Conf.,* 1984.

79. **Smith, R. E.,** Determination of chloride in sodium hydroxide and in sulfuric acid by ion chromatography, *Anal. Chem.,* 55, 1427, 1983.

80. Dionex Corporation, Method for the determination of anions in sodium hydroxide, *Appl. Note 51,* November, 1985.

81. **Wetzel, R., Smith, F. C., and Woodruff, A.,** Separation of large anions, *Dionex Corp. Appl. Note 7,* 1978.

82. **Smith, R. E.,** Applications of ion chromatography, *Nat. Tech. Info. Cent., (NTIC) Rep. No. BDX-613-3554,* 1986.

83. **Hamm, R. E.,** Complex ions of chromium EDTA complex, *J. Am. Chem. Soc.,* 75, 5670, 1953.

84. **Smith, R. E.,** Determination of Cr(III) in chromic acid, *NTIC Rep. No. BDX-613-3262,* 1986.

85. **Fitchett, A. and Woodruff, A.,** Determination of polyvalent anions by ion chromatography, *LC Mag.,* 1, 48, 1983.

86. **Smith, R. E. and Annable, P. L.,** Automated ion chromatography of plating solutions, *Am. Electr. Soc. Conf. Proc.,* in print, 1986.

87. **Merrill, R. M.,** Ion chromatographic procedure for the analysis of ethanol ammonium tartrate anodization electrolyte, *NTIC Rep. No. SAND-82-1929,* 1982.

88. Dionex Corporation, Methods development using mobile-phase ion chromatography, *Tech. Note 12R,* November 1984.

89. **Weiss, J.,** *Handbuch der Ionenchromatographie (Handbook of Ion Chromatography),* Friederich Ehrenklau Druckerei GmbH, Hessen, W. Germany, 1985.

90. **Annable, P. L. and Smith, R. E.,** Ion chromatographic analysis of alkaline soak cleaners, *Plating Surf. Finish.,* 126, March, 1986.

91. **Davis, W. D., Smith, R. E., and Yourtee, D.,** Analysis of organic additives to plating baths using an ion chromatograph, *Metal Finish.,* in print.

92. **Edwards, P. and Haak, K. K.,** A pulsed amperometric detector for ion chromatography, *Am. Lab.,* April, 1983.

93. **Westwell, A.,** Industrial applications in ion chromatography, *Anal. Proc.,* 21, 320, 1984.

94. **Pohlandt, C.,** Ion chromatography — the missing link, *S. Afr. J. Sci.,* 80, 208, 1984.

95. **Franklin, G. O.,** Development and applications of ion chromatography, *Am. Lab.,* 65, June, 1985.

96. **Littlejohn, D. and Chang, S. G.,** Determination of nitrogen-sulfur compounds by ion chromatography, *Anal. Chem.,* 58, 158, 1986.

97. **Rocklin, R. D.,** Ion chromatography: a versatile technique for the analysis of beer, *LC Mag.,* 1, 504, 1983.

98. **Dulski, T. R.,** Determination of acid concentrations in specialty alloy pickling baths, *Anal. Chem.,* 51, 1439, 1979.

99. **Cox, J. A. and Tanaka, N.,** Determination of anionic impurities in selected concentrated electrolytes by ion chromatography, *Anal. Chem.,* 57, 383, 1985.

100. **Streckert, H. H. and Epstein, B. D.,** Comparison of suppressed and nonsuppressed ion chromatography for the determination of chloride in boric acid, *Anal. Chem.,* 56, 21, 1984.

101. **Wilshire, J. P. and Brown, W. A.,** Determination of boric oxide by ion chromatography exchange, *Anal. Chem.,* 54, 1647, 1982.

102. **Schibler, J. A., Rocklin, R. D., and Rubin, R. B.,** Determination of trace anionic contaminants in sodium hydroxide by ion chromatography, *in preparation.*

104. **Eubanks, D. R., Angers, L., and Fitchett, A. W.,** Everything you ever wanted to know about boric acid analysis by ion chromatography, *Paper at 25th Rocky Mountain Conf.,* 1983.

104. **Eubanks, D. R.,** Everything you ever wanted to know about boric acid analysis by ion chromatography, II, *Paper at 28th Rocky Mountain Conf.,* 1986.

105. **Henry, L. E.,** Determination of trace cations in brines, *Paper at the 1986 Pittsburgh Conf.*

106. **Chriswell, C. D., Mroch, D. R., and Markuszewski, R.,** Determination of total sulfur by ion chromatography following peroxide oxidation in spent caustic from the chemical cleaning of coal, *Anal. Chem.,* 58, 319, 1986.

107. **Fichte, W. and Mohr, G.,** Minileckdetaktion und Ionenchromatographie (Detection of minor leaks and ion chromatography), *Maschinenschaden,* 5, 81, 1982.

108. **Plechty, M. M.,** Measurement of anions in high purity water by ion chromatography, *LC Mag.,* 2, 684, 1984.

109. **Wilshire, J. P.,** Determination of silica by ion chromatography, *LC Mag.,* 1, 290, 1983.

110. **Muller, H., Nielinger, W., and Horbach, A.,** Bestimmung von Acetylund Propionyl-Endgruppen in Poly-Epsilon-Caprolactam nach Hydrolyse mittels Ionenchromatographie (Determination of acetyl and propionyl end groups after hydrolysis with the help of ion chromatography), *Ang. Makr. Chem.,* 108, 1, 1982.

111. **Coloruotolo, J. F. and Eddy, R. S.,** Determination of chlorine, bromine, phosphorous, and sulfur in organic molecules by ion chromatography, *Anal. Chem.,* 49, 884, 1977.

112. **Wang, C. Y. and Tarter, J. G.,** Determination of halogens in organic compounds after sodium fusion, *Anal. Chem.,* 55, 1775, 1983.

114. **Fratz, D. D.,** Quantitative determination of inorganic salts in color additives, in *Ion Chromatographic Analysis of Environmental Pollutants,* Ann Arbor Press, Ann Arbor, Mich., 1977, 169.

114. **Fratz, D. D.,** Ion chromatographic analysis of food, drug, and cosmetic color additives, in *Ion Chromatographic Analysis of Environmental Pollutants,* Vol. II, Ann Arbor Press, Ann Arbor, Mich., 1978, 371.

115. **Wilson, S. A. and Gent, C. A.,** Determination of chloride in geological samples by ion chromatography, *Anal. Chim. Acta,* 148, 299, 1983.

116. **Nadkrani, R. A. and Pond, D. M.,** Applications of ion chromatography for determination of selected elements in coal and oil shale, *Anal. Chim. Acta,* 146, 261, 1983.

117. **Franklin, G. O. and Fitchett, A. W.,** Fast chemical characterization of pulping and bleaching process liquors by ion chromatography, *Pulp Paper Can.,* 83, 40, 1982.

118. **Parigi, J. S.,** Determination of sulfate in Kraft green liquor by ion chromatography, *Am. Lab.,* 124, Sept., 1984.

119. **Franklin, G. O.,** Ion chromatography provides useful analysis of the chemistry of pulping and bleaching liquors, *Tappi,* 65, 107, 1982.

120. **Weiss, J. and Gobl, M.,** Analyse anorganischer Schwefelverbindungen mit Hilfe der Ionenchromatographie (Analysis of inorganic sulfur compounds by ion chromatography), *Fresn. Z. Anal. Chem.,* 320, 439, 1985.

# INDEX

## A

Acetate, 52, 121
  in ethanol-based baths, 149
  Kraft black liquor and, 161
  in Ringer's lactate, 117
  tin and, 82
Acetic acid, 64, 86, 140
  *E. coli* and, 121
  ion exclusion and, 10
Acetonitrile, 12, 43, 96
Acetyl end group, 160
Acid copper baths, 143, 145
Acid etchant, 71, 144
Acidic tin/lead baths, 106
Acid pickling baths, 157—158
Acid rain, 2, 132
Acid zinc baths, 145
Acrylic acid, 64
Adenosine monophosphate (AMP), 56, 57
Adipic acid, 64
Aerosols, 6, 137
AFS-I, 66, 150
AFS-II, 149
AG-1 column, 42
AG-2 column, 42
AG-3 column, 42, 55
AG-4 column, 42, 43
AG-4A column, 42
AG-5 column, 42, 43
AG-6 column, 43
AG-7 column, 43
AgCl, 10, 64
Ag-form suppressor, 66
Agronomy, 122, 124, 127—130
AI 300 Data System, 108—109
Air-bags, automobile, 131, 140
Airborne inorganic acids, 138
Airborne particulates, 6, 132, 137
Air pollution, 137—138
Air samples, 131
Alcohols, 155
  in beer, 157
  electrochemical detection of, 18
  pulsed amperometric detection of, 91
  in vinegar, 121
Aliphatic amines, 19—20, 71
  column deterioration and, 42—43
  in plating solutions, 152
Aliphatic carboxylic acids, 149
Alkaline cleaning solution, 141, 153
Alkaline earth metal cations, 20, 72
  column contamination and, 43
  in urine, 116
Alkaline hydrogen peroxide impinger system, 140
Alkaline plating solutions, 106
Alkanol amines, 79

Alkyl detergents, 152
Alkyl phenol polyethoxylates, 105, 153
Alkyl phenones, 106
Alkyl sulfates, 36
Alkyl sulfonates, 95, 102
Aluminum, 78
Aluminum industry, 138, 140
American Society for Testing and Materials
      (ASTM), 133
  rain water analysis procedures of, 6
Amines, see also specific amine; specific type of
      amine
  in plating solutions, 152
  primary, *o*-phthaldehyde and, 66
  secondary, 85
Amino acids, 1, 3, 85—86, 121—122
  HPICE AS-4 and, 66
Amino groups, quaternary, 9
*p*-Aminophenyl arsonic acid, 134
Ammonia, 79
Ammonium, 154
  HCl and, 4
  in urine, 116
Ammonium bifluoride/HCl etchant, 145
Ammonium hydroxide, 96
Ammonium plant, steam-water cycle of, 155
AMP (adenosine monophosphate), 56, 57
Aniline, 155
Anion(s)
  aromatic, MPIC and, 96—97
  common, 48—63, see also specific anion
    AG-3 and, 55
    AS-3 and, 54
    AS-4 and, 56
    AS-5 and, 58—60
    AS-6 and, 61—63
    Auto-Ion System 12 and, 56
    brine column and, 53
    HPIC and, 56
    MFC-1 and, 63
    Nafion 811 and, 58
    Series 2000 and, 56
  counter, 101
  hydrophobic
    ICE and, 10
    MPIC and, 12
  standard seven, HPIC and, 11
Anion detection
  electrochemical detection and, simultaneous, 89—
      90
  metal determinations and, simultaneous, 84
  without chemically suppressed conductivity, 18
Anion fiber suppressor, 11—12, 66
  plating solution analysis with, 149
Anion micromembrane suppressor, see Micromem-
      brane suppressor
Anion separator, see AS

Anodic peak, 91
Antarctica, snow samples from, 132
Anticarcinogen, oxalate as, 118
AOAC (Association of Official Analytical Chemists), 128
APM (atmospheric-particulate material), 138
Apple juice, 128
Applications, 115—162
    environmental sciences, 131—142
    life sciences, 115—131
    other, 155—162
    plating solution analysis, 141—154
Arginine, 86
Aromatic amines, 19, 22, 115
    column deterioration and, 42—43
    environmental, 131
    MPIC and, 102, 104
    oxidation of, 104
    in plating solutions, 152
Aromatic anions, MPIC and, 96—97
Aromatic-aromatic interaction, 105—106
Aromatic rings, column deterioration and, 42
Aromatic sulfonates, 96
Arsenate, 54, 134
Arsenic, 54
Arsenite, 54, 134
Arsine derivatives, 134
Aryl sulfonates, 95
AS-1 column, 5, 147, 160
AS-2 column, 7, 10
AS-3 column, 11, 54, 117, 118
AS-4 column, 11, 56, 90, 147
AS-4A column, 117, 138
    beer analysis and, 156
    physiological fluids and, 121
    plating solution analysis and, 147
AS-5 column, 7, 58—60, 117
    automation and, 107
    physiological fluids and, 121
    plating solution analysis and, 144
AS-6 column, 61—63, 92
    detergents and, 161
    physiological fluids and, 121
    sugars and, 121
AS-7 column, 18, 85, 149
AS-8 column, 23, 86, 122
Ascorbate, 70
Ascorbic acid, 66
    urinary, 118
Aspartic acid, 86
Association of Official Analytical Chemists (AOAC), 128
ASTM, see American Society of Testing and Materials
Atmospheric-particulate material (APM), 138
Atmospheric samples, 131, 132, 137
Atomic absorption spectroscopy, 77
    urine analysis with, 116
    water analysis with, 134
$Au^+$, 85
$Au^{2+}$, 150

$Au^{3+}$, 85
AuCl, 60
$AuCl_4^-$, 85
$Au(CN)_2^-$, 44, 85, 90, 150—151
$Au(CN)_4^-$, 85
Auto-Ion System 12, 56
Auto-Ion 300 Data System, 108—109
Automation, 24, 106—112
    anion fiber suppressor and, 12
    conductivity cell and, 108
    different analyses with, 109—110
    module used in, 107
    power plants and, 106—107
    in pulp and paper industry, 161
    Series 2000 and, 108—109
    Series 8000 and, 110—112
    suppressor used in, 108
    water dip and, 133
Automobile air-bags, 131, 140
Autosampler, 24
    activation of, 109
    plating solutions and, 143, 154
A valve, 24—25, 107
Aza arenes, 115
    environmental, 131
Azide, 115, 131, 140
Azo-sulfonated organic dyes, 160

**B**

Bacon, nitrite levels in, 128
Bacteria, sulfate-reducing, 121
Barium, 19, 71, 154
Beans, green, 128
Beef, 127
Beer analysis, 2, 155—157
Benzene sulfonate, 96—97
Benzene sulfonate detergents, in plating solutions, 152
Benzoic acids, 10, 57
Beverages, flavor of, 115
Bio-pond waters, 134
Birmingham, Alabama, sulfur dioxide measurements in, 138
Bismuth, 78
Bisulfite, 136
Bleach, 161
    hypochlorite in, 160
Blood serum, lithium in, 117
Boiler-water composition, 155
Bombing investigations, monomethylamine determination in, 71
Borate, 16
Borate-mannitol complex, 149
Boric acid, 86, 159
    gradient elution and, 36
    micromembrane suppressor and, 13
    in plating solutions, 149
    urinary oxalate determination and, 118
Bread, potassium bromate in, 124
Brewery, see Beer analysis

Brighteners, 147, 152
Brine column, 10, 53
Brines, 128, 159
Bromate, 52, 160
Bromide, 4, 87, 90
  electrochemical detection of, 18
  ion exclusion and, 10
*m*-Bromo benzoic acid, 57
Butanophenone, 106
Butyric acid, 64
B valve, 25, 108
BWR Owners Group, 134

# C

CAC (chloroacetyl chloride), 140
Cadmium, 143
Calcium, 19, 71, 78, 154
  in brines, 159
  in plating baths, 149
Capacity factor, 25, 101—102
Carbohydrates, 2, 61
  in beer, 155—156
  electrochemical detection of, 90
  working electrode and, 22
Carbonate, 79, 121, 160
  in gold cyanide bath, 149
Carbon electrode, glassy, 18, 104, 153
Carbon treatment, plating solutions and, 152
Carboxylic acids, 10
Carboxymethyl, 7
Carrots, nitrates in, 127
Catecholamines, oxidation of, 104
Catechols, 18, 153
Cathodic peak, 91
Cation(s)
  divalent, 20
    column contamination and, 43
  inorganic, aliphatic amines vs., 152
  ion-pairing, 101
Cation analysis, 71—73
  advances in, 19—20
  in plating solutions, 154
Cation exchange resin, 3
Cation exchange site, 66
Cation fiber suppressor, 108
Cation micromembrane suppressor (CMMS), 19—
    20, 102
  beer analysis and, 157
  urine analysis and, 116
Cation separator, see CS
Caustic, spent, total sulfur in, 159
$Cd^{2+}$, 77
Cd(CN), 90
Cd-EDTA, 84
Cellulose, 7
Cerebrospinal fluid, 117
Cetylpyridinium, in mouthwash, 102
Cetyltrimethyl ammonium bromide (CTAB), 78
CFIA (Committee of Food Industry Analysts), 127
CG-1 column, 43

CG-3 column, 43
Charcoal filter, 140
Charcoal sorbent, sulfur dioxide and, 138
Chelating agents, 78
  ferric nitrate and, 18
  transition metals and, 19
Chelex, 52, 148
Chemical mutagens, 115
Chemical suppression of conductivity, 1, 4—6, see
    also Conductivity
Chlor-alkali membrane process, 159
Chlorate, 52
Chloride, 4, 6
  ion exclusion and, 10
  in plant materials, 122, 124
Chloroacetate, 160
Chloroacetyl chloride (CAC), 140
Chlorobenzoic acid, 26—27, 57
Chromate, 12, 115, 134, 155
  column performance and, 42
  in plating solutions, 148
Chromatography module, 24
Chrome azurol S, 78
Chromic acid, 2, 39
Chromium, 155
Cider vinegar, 121
Circuits
  integrated, 160
  printed, 155
Citrate, 42, 149
Citric acid, 19, 64, 66, 71, 128, 142
  transition metals and, 77
Cloud water, see Fog water
CMMS, see Cation micromembrane suppressor
CMP (cytidine monophosphate), 56, 57
$Co^{2+}$, 78, 81, 82, 85
Coal, 138
  sulfur-containing, 137
Coal fly-ash, 137
Coal gasifier, fixed-bed, 134, 136
Cobalt, 143, 144
Cobalt cyanide complex, 144
Cobaltous ion, 150—151
Co(CN), 85, 90
Co-EDTA, 84
Coffee, 107
Color additives, 160
Column, see specific type
Column care, 41—44
Column efficiency, 27
Column gradient constant, 33
Column performance, 41—44
Column selectivity, see Selectivity
Commercial shampoos, 161
Committee of Food Industry Analysts (CFIA), 127
Compositional gradient, 30
Computer-interface module, 110
Computers, see also Automation; Microprocessor
  printed circuits in, 155
Concentration gradient, 29
Concentrator module, 107

Conductance, 40—41
Conductivity, 37—41, see also Electrochemical
    detection
  chemical suppression of, 1, 4—6
Conductivity cell, 108
Conductivity detection, 4, 47—48
Copolymerization, suspension, 7
Copper, 3, 82, 143
Copper bath, electroless, 149
Copper-dihydroxybenzene complexes, 153
Copper sulfate plating bath, 144
Copper sulfate/sulfuric acid bath, brighteners in, 152
Core particle, AS—3, 54
Corn, canned, 127
Corn vinegar, 121
Corrosion inhibitors, 72
Corrosion resistance, 141
Corrosive plating solutions, organic additives to, 105
Counter anion, 101
Counter electrode, 21
  platinum, 89, 134
Cr$^{3+}$, 81, 148
Cr(VI), 134
Cr-PDCA, 81, 148
Creatine phosphate, 52
Cross-linking, AS—3, 54
CS-1 column, 5, 71—72
  anion retention by, 43
  A valve and, 108
  plating solutions and, 154
  urine analysis and, 116
CS-2 column, 19, 72, 128, 143
CS-3 column, 19, 72—73
  beer analysis and, 157
  plating solutions and, 154
CS-4 column, 19, 86
CS-5 column, 19, 78—79, 81
  transition metals and, 145
CSF (cerebrospinal fluid), 117
CTAB (cetyltrimethyl ammonium bromide), 78
Cu$^{2+}$, 77, 78, 82
Cu(CN), 90
Cu-EDTA, 84
Cyanide, 87, 89
  column performance and, 42
  electrochemical detection of, 18
  environmental, 136
  MPIC and, 95
4-Cyanophenol, 147
*p*-Cyanophenol, 13, 58, 136
Cyclic voltametry, 91
Cyclohexylamine, 72
Cytidine monophosphate (CMP), 56, 57

**D**

Debye-Huckel-Onsager equation, 39—40
Dehumidifier, air samples and, 138
Deionized water, ultrapure, 131
Depression, lithium and, 117
Dequest agents, 150

  phosphonates as, 149
Desulfovibrio, 121
Detergents, 10, 96, 97, 155, 160—161
  nonionic, 106
  in plating solutions, 151
Dextran, 7
Diaminopropionic acid, 73
Dibutylphosphoric acid, 51
Dicarboxylic acids, 121
Dichromate, 155
Diesel exhaust fumes, 131
Diet Coke, 70
Diethanolamine, 72, 152
Diethylamine, 72
Diethylaminoethane, 7
Diethylenetetramine pentaacetic acid, see DTPA
Dilution module, 107
Dimethyl arsinic acid (DMA), 134
2,2-Dimethyl propionic acid, 64
Dionex Corp., 2
Diprotic acids, 71
Disaccharides, 18, 92—93
Distribution coefficient, 25
Disulfide anionic brightener, 152
Dithionate, 90, 98
Divalent cations, 20
  column contamination and, 43
  in plating solutions, 154
Divinylbenzene, 7, see also Polystyrene/divinylben-
    zene
DMA (dimethyl arsinic acid), 134
Dodecylsulfate, 96
Donnan exclusion, 6, 10, 64
  organic anions in plating baths and, 149
Dow Chemical Co., 2
Dowex, 10, 54
Drinking water, 6, 134, see also Water analysis
  Safe Drinking Water Act and, 132
DTPA (diethylenetetramine pentaacetic acid), 18, 85
Dyes, organic, 155, 160
Dynamic regenerant concentration, 16
Dynamic suppression capacity, 13—15, 29, 159

**E**

EDTA (ethylenediamine tetraacetic acid), 2, 60, 82,
    144, 148
Effective capacity factor, 32
Effluents, 6
Electric double layer, 99—100
Electric power plants, 155
Electric utilities, 134
Electrochemical detection, 2, 18, 22, 86—87, 89—95
  of carbohydrates, 92—95
  of cyanide, 87, 89, 136
  in pulp and paper industry, 161
  pulsed amperometric detector and, 90—93
  simultaneous multi-anion analysis and, 89—90
Electrodes, see specific type
Electroless copper bath, 149
Electroless nickel bath, 144, 145, 149

Electronics industry, 141
Electroplating industry, 141
Electrostatic bond, 9
EMPA (ethyl methyl phosphonic acid), 51
Environmental analysis, 131—142, see also specific
    application
Environmental chemists, 131
Environmental genotoxicants, 115
Environmental Protection Agency (EPA), 131, 133
  procedures of, 6
  water analysis by, 1
Equivalent conductance, 38
*Escherichia coli*, in fermentation medium, 121
Ethanol, in vinegar, 121
Ethanolamines, 72, 131, 140
Ethanol-based baths, 149
Ethanophenone, 106
Ethylamine, 72
Ethylenediamine, 22, 72, 78, 90
  in veterinary livestock iodine supplement, 124
Ethylenediamine tetraacetic acid (EDTA), 2, 60, 82,
    144, 148
Ethyl methyl phosphonic acid (EMPA), 51
Exhaust fumes, 131

### F

Fast-run column, 10
$Fe^{2+}$, 82
$Fe^{3+}$, 77, 78, 81
Fe(CN), 90
Federal Republic of Germany, environmental
    analysis in, 132
Fermentation, 69
  beer, 155
  *E. coli* and, 121
  vinegar and, 121
Ferric chloride, 54
Ferric nitrate, 18, 149
Ferrous ion, 142
Fertilizer, urea in, 124
Fiber suppressor, 11— 12, see also Anion fiber
    suppressor; Cation fiber suppressor
Fixed-bed coal gasifier, 134, 136
Flame photometry, lithium determination with, 117
Flash-distillation, sulfite extraction with, 129
Flavor, food and beverage, 115
Flue-gas scrubbing liquor, 155
Fluorescence detectors, 2, 122, see also Post-column
    reaction
  amines and, 152
Fluoride, 4
3-Fluorobenzoic acid, 104
*m*-Fluorobenzoic acid, 57
Fluoroborate, 145
Fluoroboric acid, 153
Fluorophor, 122
Fluoropore FA membrane filters, 138
Fog water, 136—137
  hydroxymethanesulfonate in, 1
Foods

flavor of, 115
  sulfite in, 129
Forest areas, 132
Formaldehyde
  in fog water, 136
  sulfite and, 52, 129
Formate, 52, 138
  in ethanol-based baths, 149
  Kraft black liquor and, 161
  in Ringer's lactate, 117
Formic acid, 10, 64, 140
Fructose, 121
Fructose-1,6-diphosphate, 52
Fructose-6-phosphate, 52
Fuel oils, 138
Fumarate, isopropanol and, 68—69
Fumaric acid, 64
Fumigant, 140

### G

GaAs crystal, 81
Galacturonic acid, 66
Gallium, 81
Gas-diffusion denuders, air samples and, 138
Gas streams, 137
Genotoxicants, environmental, 115
Geological samples, 134, 161
Geothermal wells, 161
Germany, environmental analysis in, 132
Glassy carbon electrode, 18, 104, 153
Glucose, 91, 121
Glucose-1-phosphate, 52
Glucose-6-phosphate, 52, 60
Glucuronic acid, 66
Glutamic acid, 86
Glyceric acid, 64
Glycerol, 121
Glycerol-3-phosphate, 52
Glycine, 13
Glycolate, 160
Glycols, electrochemical detection of, 18
Glycosidic bonds, 156
GMP (guanidine monophosphate), 56
Gold, 60, 84—85
Gold baths, 149, 150
Gold cyanide baths, 144, 149
Gold electrode, 22, 90
Gold oxide, 91
Gout, oxalate and, 118
Gouy layer, 100
Gradient constant, 32
Gradient elution, 6, 29—37, 155
  alkyl sulfates and, 36
  boric acid and, 36
  column gradient constant and, 33
  compositional gradient and, 30
  concentration gradient and, 29
  dynamic suppression capacity and, 29
  effective capacity factor and, 32
  gradient constant and, 32

instantaneous capacity factor and, 32
isocratic column constant and, 33
mannitol and, 36
microprocessor and, 107
ramp rate and, 32
sulfonates and, 36
Green beans, 128
Groundwaters, 55
Guanidine monophosphate (GMP), 56
Guanosine, 57
Guard columns, 10, 115
  deterioration and, 41

**H**

Halides, 160
Ham, nitrates in, 127
HCl, 4
HCN, 136
Heart-cutting, 55
Height equivalent per theoretical plate (HETP), 28
Hexafluorosilicate, 160
Hexane sulfonic acid, 99, 102
Hexanophenone, 106
Hexyl sulfonates, 96
$Hg^{2+}$, 79, 82
High-performance ion chromatography (HPIC), 11,
  56
High-performance ion chromatography exclusion
  (HPICE), 65—67, see also Ion chromatogra-
  phy exclusion
High-purity NaOH, 147
Hippuric acid, 57
Histidine, 86
HMSA, 136—137
Hollow fiber, 11—12
HPIC, see High-performance ion chromatography
HPICE, see High-performance ion chromatography
  exclusion
HPICE AS—1, 66, 121, 149
HPICE AS—2, 66
HPICE AS—3, 66
HPICE AS—4, 66
HPICE AS—5, 66, 121
$HSO_3^-$, in fog water, 136
Hull cell, 152
Human plasma analysis, 117
Humic acids, 13, 41
Hybrid microcircuits, 141
Hydrazide, environmental, 131
Hydrazine, 18, 72, 115
Hydrogen bonding, HPICE AS-5 and, 66
Hydrogen cyanide, 138
Hydrogen peroxide, 159—160
Hydrophobic anions
  ion exclusion and, 10
  metal-chloride complexes, 60
  MPIC and, 12
  in plating solutions, 149
Hydrophobic interactions, 66
Hydroquinone, in plating solutions, 153

Hydroxy acids, 66
α-Hydroxy butyric acid, 64
Hydroxyimidosulfate, 155
2-Hydroxyisobutyric acid, 82
Hydroxylamine, environmental, 131
Hydroxylamine hydrochloride, 128
Hydroxymethanesulfonate, 1, 136
Hydroxysulfamate, 155
Hypochlorite 18, 160
Hypophosphite, 145

**I**

ICE, see Ion chromatography exclusion
IC/ICE (ion chromatography/ion chromatography
  exclusion), 120
ICP (inductively coupled plasma) spectroscopy, 77
Igepal, 105, 153
Imidodisulfate, 155
Iminodiacetate, 159
IMPA (isopropyl methyl phosphonic acid), 51
Independent migration, law of, 38
Inductively coupled plasma (ICP) spectroscopy, 77
Industrial hygiene, 138, 140
Industrial waste, 134
Inorganic acids, airborne, 138
Inorganic cations, aliphatic amines vs., 152
Inorganic modifier, 98
Inositol-2-phosphate, 60
Instantaneous capacity factor, 32
Integrated circuits, 160
Intravenous solution, amino acids in, 122
Iodate, 52, 160
Iodide, 58, 90, 160
  electrochemical detection of, 18
  in milk, 129
Iodine supplement, veterinary livestock, ethylenedi-
  amine in, 124
Ion chromatography exclusion (ICE), 10, 64—71, 86
  alcohols and, 92
  anion fiber suppressors and, 66
  AS-5 and, 69—70
  automation and, 107
  beer analysis and, 157
  columns for, performance of, 43
  Donnan exclusion and, 64
  high-performance, 65—67
  hydroxy acids and, 66
  metabolites and, 71
  organic acids in physiological fluids and, 120
  organic anions in plating baths and, 149
  packed-bed suppressor and, 64
  pKa and, 68—69
Ion chromatography/ion chromatography exclusion
  (IC/ICE), 120
Ion exchange, 3
Ion exchange capacity, 2
Ion exchange equivalents, 2
Ion-ion interactions, 39
Ion pair, 96
Ion pair chromatography, 6

Ion-pairing cation, 101
Ion-pair reagent, 12, 96—98
Iron, 78, 128
Irrigation waters, 122
Isobutyric acid, 64
Isocitrate, 70
Isocitric acid, 66
Isocratic column constant, 33
Isoelectric point, 15
Isopropanol, 68—69
Isopropyl methyl phosphonic acid (IMPA), 51

**K**

α-Ketobutyric acid, 64
α-Ketovaleric acid, 64
Kidney failure patients, monitoring of, 117—118
Kidney stones, oxalate and, 118
KOH, 61, 121
Kohlrausch Law, 40
Kraft black liquor, 161
Kraft smelt, 161

**L**

Lactate, 120, 121, 161
Lactic acid, 64, 71, 121
Lakes, pH of, 132
Landfill leachate, 136
Lanthanide metals, 76
Latex, 9, 42, 60
Laundry detergent, 97
Law of independent migration, 38
Lead, 60
  EDTA and, 82
  in plating solutions, 143, 144
Limiting equivalent conductance, 38
Linear alkyl detergents, 152
Linear velocity, 28
LiOH, 77, 82, 142
Liquid laundry detergent, 97
Lithium, 116—117, 154
Lithium perchlorate, 104
Liver, analysis of, 117
Livestock iodine supplement, ethylenediamine in, 124
Lysine, 86

**M**

Machine cutting fluids, 155
Macroporous resins, 20
Magnesium, 19, 71, 78, 154
  in brines, 159
  in plating baths, 149
Maleate, 57
Maleic acid, 64
Malic acid, 66
Malonate, 57
Malonic acid, 64
Mandelic acid, 57, 64

Mannitol, 36, 149, 159
MAPS-3 region, acid rain samples from, 132
Membrane filters, Fluoropore FA, 138
Mercury, 82
Metabolites, 71
Metal-cutting fluids, nitrites in, 140
Metal-cyanide complex, 90, 95, 136
  column performance and, 42
  micromembrane suppressor and, 17
Metal determinations, anion determinations and, simultaneous, 84
Metal-EDTA complexes, 60, 144, 148
Metal-finishing water, 136
Metal-free column, 63, 84
Metal processing, 141
Methanol, 71—72, 86, 127
  column performance and, 43
Methylamine, 72
o-Methyl hippuric acid, 57
p-Methyl hippuric acid, 57
Methyl phosphonic acid (MPA), 51
MFC-1, 63
Microcircuits, hybrid, 141
Micromembrane suppressor (MMS), 7, 13—18, 159
  cation, see Cation micromembrane suppressor
  ICE, 67
  organic solvents and, 43
  plating solution analysis and, 148, 149
  suppression mechanisms of, 14
Microprocessor, 24, 107, see also Automation
  plating solution analysis and, 147
  water dip and, 133
Milk, 107
  acetic acids in, 71
  iodide in, 129
Minerals, precious, 155
Mining-process streams and effluents, 155
MMA (monomethyl arsonic acid), 134
MMS, see Micromembrane suppressor
$Mn^{2+}$, 77, 78
Mn-EDTA, 84
Mobile-phase ion chromatography (MPIC), 12, 20—21, 95—106
  aromatic amines and, 102, 104
  capacity factor and, 101—102
  columns for, performance of, 43—44
  detergents and, 105—106
  electric double layer and, 99—100
  electrochemical detection and, 104
  environmental uses for, 138, 140—141
  inorganic modifiers in, 98
  iodide and, 129
  ion-pair reagent in, 96—98
  metal-cyanide complex determination with, 90
  nonionic compounds and, 104—105
  plating solution analysis with, 150—154
  reversed-phase HPLC vs., 95—96
  UV-visible detector and, 104
Molybdate, 12, 42, 52
Molybdenum, 52
Monocarboxylic acids, 121

Monoclonal antibodies, 115
Monoethanolamine, 102
Monomethylamine, 71
Monomethyl arsonic acid (MMA), 134
Monosaccharides, 18, 92
Monovalent cations, 154
Morpholine, 72
Mouthwash, cetylpyridinium in, 102
MPA (methyl phosphonic acid), 51
MPIC, see Mobile-phase ion chromatography
Multi-anion analysis, simultaneous, electrochemical
     detection and, 89—90
Multicomponent determinations, 131
Mutagens, chemical, 115

**N**

Nafion 811, 58
NaOH, 86, 147, 157
National Food Processors Association, 127
National Interim Primary Drinking Water Regula-
     tions, 55
$Ni^{2+}$, 77, 78
Ni(CN), 90
Ni-EDTA, 84
Nickel, 82, 143, 145
Nickel bath, electroless, 144, 145, 149
Nickel sulfamate bath, 145
Ni/Fe bath, 149
Ninhydrin, 1, 85
Nitrate, 4, 6, 10
Nitric acid, 8, 85
Nitridotrisulfate, 155
Nitrilotriacetic acid, 85
Nitrites, 115
   anion fiber suppressor and, 12
   environmental, 131
   in metal-cutting fluids, 140
   MPIC columns and, 44
Nitrobenzenes, 18, 21, 104
Nitrosamines, 132
*N*-Nitrosohydroxylamine-*N*-sulfonate, 155
Nonionic detergents, 106
Nonionic organic additives, 152—153
Nuclear power plants, 134
   automation and, 106, 111
Nucleosides, 57
   monophosphates, 121
   triphosphates, 54, 121
Nucleotides, 54, 56

**O**

Oatmeal, 127
Ocean waters, 134
Octane sulfonic acid, 16, 67, 102, 140
Oil shales, 161
Oligosaccharides, 92—93
   in beer, 156
   electrochemical detection of, 18
Organic additives, to plating solutions, 105, 147,
    152—153
Organic cleaner, treated, 136
Organic dyes, 155, 160
Orthopolyphosphate, 85
Oxalate, 52, 57
   urinary, 118, 120
Oxalic acid, 19, 64, 128, 142
   transition metals and, 77
Oxyanions, 12
Oxygen scavenger, 72

**P**

Packed-bed suppressor, 11, 55, 64, 67
   plating solution analysis and, 149
Palladium, 84—85
Paper industry, 155, 161
PAR, see 4-Pyridyl azo resorcinol
Parr bomb, 138, 160
Particulates, 6, 132, 137
$Pb^{2+}$, 78
Pb-EDTA, 84, 148
PDCA (pyridine-2,6-dicarboxylic acid), 78, 81, 148
$PdCi$, 60, 85
Pellicular resins, 1
Pentanophenone, 106
Pentyl sulfonates, 96
Perchloric acid, 55, 104, 138
Perfluorobutyric acid, 66
Peroxide-absorbing solution, 137
Persulfate, electrochemical detection of, 18
Phenols
   electrochemical detection of, 18
   MPIC and, 21
   oxidation of, 104
Phenyl, sulfuric acid and, 7
Phenylalanine, 86
*m*-Phenylenediamine, 19, 71, 133, 154
Phenylthiohydantoin (PTH), 85
Phosphate/citrate-buffered gold plating bath, 151
Phosphates, 4, 6, 79, 145, 160
   sugar, 7, 58, 60, 115
Phosphite, 145
Phosphoenol pyruvate, 52
Phosphonates, 18, 149
Phosphonic acids, 51
Phosphoric acid, 10, 56
Photometry, flame, lithium determination with, 117
*o*-Phthaldehyde, 18, 85, 152
   amino acids and, 122
   primary amines and, 66
Physiological fluids, see also specific determination;
    specific fluid, 115
   organic acids in, 120
   sample preparation with, 121
Phytic acid, 55
Pickle liquor, treated, 136
Pickling baths, 157—158
pKa, ion exclusion and, 10, 68—69
Plant effluents, 6
Plant hydrolysates, monosaccharides in, 92

Plants, 122, 124
Plasma analysis, 117
Plating solutions, 141—154
  alkaline, 106
  AS-7 and, 149
  column choice and, 147—148
  column performance and, 41
  conductivity of, 39
  CS-5 and, 144
  electroplated metals and, 142
  ICE and, 149—150
  micromembrane suppressor and, 146—147
  MPIC and, 150—154
  organic additives to, 105, 147
  tin-lead alloy, 82
  transition metals and, 144—145
Platinum, 60, 84—85
Platinum electrode, 18, 89, 92, 134
Pollution
  air, 137—138
  potential sources of, 131
Polyamines, 20, 152
Polycarboxylic acids, 85
Poly-epsilon-caprolactam, 160
Polymeric resins, 3
Polymerization, degree of, carbohydrates and, 156
Polyphosphates, 18
Polyphosphonates, 85, 161
Polysaccharides
  in beer, 156
  pulsed amperometric detector and, 22
Polystyrene/divinylbenzene, 3, 7, 95
  column deterioration and, 42
Poragens, 95
Post-column reaction, 4—6, 75—89
  amino acid determination with, 85—86
  CS-5 and, 78—79
  EDTA-metal complexes and, 82—84
  gallium and, 81
  mercury and, 82
  PAR formation kinetics and, 79—81
  polyvalent anion determination with, 85
  spectrophotometric reagents and, 78
  tin and, 82
  transition metals and, 76—78
  vanadium and, 81
Potassium, 4, 116, 154
Potassium bromate, 124
Potassium dichromate baths, 148
Potassium gold cyanide, 150
Potato chips, 121
Potentiostat, 89
Power plants, 72, 134, 155, 160
  automation and, 106—107, 111
Precious minerals, 155
Precipitation, 6
Preconcentration, 159
Printed circuits, 155
Proline, 18, 85
Propanophenone, 106
Propionic acid, 64

Propionyl end group, 160
Proteins, in urine, 116
PtCl, 60, 85
PTH (phenylthiohydantoin), 85
Pulp and paper industry, 155, 161
Pulsed amperometric detector, 18, 22, 76, 86, 90—93
  beer analysis and, 156
Pump module, 107
Pyridine-2-carboxylic acid, 19
Pyridine-2,6-dicarboxylic acid (PDCA), 78, 81, 148
4-Pyridyl azo resorcinol (PAR), 76, 128, 142, 155
  formation kinetics of, 79—81
  post-column reaction with, 5, 19
Pyropolyphosphate, 85
Pyruvate, 70, 160
  in Ringer's lactate, 117
Pyruvic acid, 64, 71

## Q

Quaternary amines, 9, 95
Quinic acid, 66

## R

Rain water, 132
  ASTM procedure for analysis of, 6
Ramp rate, 32
Reference electrode, 21, 134
  amperometric flow-through cell, 89
Refinery water, 141
Regenerant, 11, 16
Renal failure patients, monitoring of, 117—118
Replenishing solution, 152
Resolution, 28
Resorcinol, 153
Reversed-phase ion pair chromatography, 20
Ringer's lactate, 117
Roadside samples, 137

## S

Saccharides, 92
Safe Drinking Water Act, 132
Salicylic acid, 57
Sample-collection devices, 137
Sample preparation, 121
Schoniger flask, 122, 138, 160
Scrubber water, 136
Scrubbing vessel, 140
SDWA (Safe Drinking Water Act), 132
Selectivity
  definition of, 26
  equation for, 27
Selenate, 52, 134
Selenite, 52, 134
Selenium, 52, 134
Semiconductors, gallium and, 81
Separator columns, 2, 3
Sequoyah Nuclear Plant, 134
Series 2000, 56, 108—109

plating solutions and, 154
Series 8000, 110—112
    plating solutions and, 143, 154
Serum, lithium in, 117
Shampoos, 161
Shellfish, 127
Silanizing agents, 67
Silica, 160
    MPIC and, 12
Silica gel absorber, 138
Silicate rocks, 161
Silver, 10
Silver electrode, 18, 21, 134
    cyanide and, 87, 89, 136
Silver form, packed-bed suppressor in, 67
Sludges, cyanide in, 136
Sn(II), 82
Sn(IV), 82
Snow, 132
-SO$_3$H, 99
Sodium, 4, 116, 154
Sodium bicarbonate, 11
Sodium borate, 90
Sodium borohydride, 134
Sodium carbonate, 11, 90, 98
Sodium perchlorate, 60
Sodium phosphate, 56, 104
Sodium tartarate, 43
Sodium tetraborate, 16, 51
Soft drinks, sweetened, 121
Soil, 107, 122, 131, 161
    water washings of, 115
Solenoid-activated switching, 107
Solvation sphere, 40
Soy flour, sugars in, 92
Specific conductance, 38, 86
Spectroscopy, atomic, 77
    urine analysis with, 116
    water analysis with, 134
Standard seven anions, HPIC and, 11
Stannous ion, 82
Steam-driven turbines, in power plants, 106—107
Steam Generator Owners Group, 134
Steam-water cycle, ammonium plant, 155
Steel etchant, 144
Stern layer, 100
Streams, pH of, 132
Stripper column, 4, see also Suppressor column(s)
Strong cation exchange site, 66
Strontium, 19, 71, 154
Styrene, 7, see also Polystyrene/divinylbenzene
Succinate, 57, 70, 120, 149
Succinic acid, 64, 121
Sucrose, 121
Sugar alcohols, 18, 22, 92
Sugar phosphates, 7, 58, 60, 115
Sugars
    in beer, 156
    pulsed amperometric detection of, 91
    in vinegar, 121
Sulfamate, 52, 145, 155

Sulfate, 4, 6, 121, 160
    ion exclusion and, 10
Sulfate-reducing bacteria, 121
Sulfide, 18, 87, 90
Sulfite, 18, 52, 129
Sulfonated polystyrene/divinylbenzene, 3, 7, 42, 95
Sulfonates, 3
    aromatic, 96
    gradient elution and, 36
    MPIC and, 95
Sulfonation layer, depth of, 8
Sulfosalicylic acid, 78
Sulfur
    in coal, 137
    in plant materials, 122, 124
    in spent caustic, 159
Sulfur dioxide, 115
    atmospheric, 137—138
    environmental, 131
Sulfuric acid, 7, 39
Sulfuric acid aerosols, 137
Sulfuric-acid-anodizing bath, 141, 146
Sulfuryl fluoride, 140
Suppressor column(s), 2, 4, 5, see also specific type
Surface-active agents, in plating solutions, 151
Suspension copolymerization, 7
Sweetened soft drinks, 121
Sweet wort, 155

# T

Tartarate, 43, 57, 149
Tartaric acid, 64, 121, 142
TBAOH (tetrabutyl ammonium hydroxide), 67, 96, 129, 151
Teflon filters, 137
Tennessee Valley Authority, 134
Termite control fumigant, 140
Tetrabutyl ammonium chloride, 137
Tetrabutyl ammonium hydroxide (TBAOH), 67, 96, 129, 151
Tetradecylsulfate, 96
Tetrahydrofuran, 17
Tetrapropyl ammonium hydroxide (TPAOH), 12, 44, 150
    thiocyanate and, 96
Tetrathionate, 98
Theoretical plate, 26
    column efficiency and, 27
    height equivalent per, 28
Thiocyanate, 58, 90
    electrochemical detection of, 18, 21
    TPAOH and, 96
    in water, 136
Thiosulfate, 58, 98, 136
Three-electrode cell, 21
Tin/lead baths, 82, 153
    acidic, 106
    fluoroborate, 145
*m*-Toluidine, 104
Tomatoes, canned, 127

Tomato juice, 70, 127
TPAOH, see Tetrapropyl ammonium hydroxide
Transition metals, 4, 76—78, 141, see also specific
    metal
  chelating agents and, 19
Treated organic cleaner, 136
Treated pickle liquor, 136
Tributylphosphoric acid, 51
Triethanolamine, 72, 152
Triethylamine, 72
Trimethylamine, 72
Tripolyphosphate, 85
Triton X-100, 78, 105, 161
  plating solution analysis and, 153
Tungstate, 12, 52
Tungsten, 52
Turbine blades, 160
  chloride and, 134
Turbines, steam-driven, in power plants, 106—107
Two-dimensional chromatography, 55
  organic acids in physiological fluids and, 120
Tyrosine, micromembrane suppressor and, 15

## U

Ultrapure deionized water, 131
UMP (uridine monophosphate), 56
University of North Dakota Energy Research Center,
    134, 136
Urea, in fertilizer, 124
Urease, 124
Uridine, 57
Uridine monophosphate (UMP), 56
Urine analysis, 116
  oxalate determination in, 118, 120
  VMA in, 120
U.S. EPA, see Environmental Protection Agency
UV-visible detectors, 2, 18, see also Post-column
    reaction
  chromate and, 134

## V

$V^{4+}$, 81
$V^{5+}$, 81
Vanadium, 81
Vanillylmandelic acid (VMA), 120

Vapors, 132
Veterinary livestock iodine supplement, ethylenedi-
    amine in, 124
Vinegar, 121
Vitamins, water-soluble, 128
VMA (vanillylmandelic acid), in physiological
    fluids, 120
Void volume, 14, 25

## W

Waste-treatment sludge, 136
Wastewater, 6, 52, 131, 136
Water, deionized, ultrapure, 131
Water analysis, 6, 134, 155
  boiler, 155
  EPA methods for, 1
  irrigation, 122
  refinery, 141
  Safe Drinking Water Act and, 132
Water dip, 12, 133, 146
Water rinses, 145
Water-soluble vitamins, 128
Water washings, of soil samples, 115
Watt's nickle, 149
Weak cation exchange site, 66
Wine, 155
  diprotic acids in, 71
Working electrodes, 21
  amperometric flow-through cell, 89
  carbohydrates and, 22

## X

Xylene sulfonate, 97

## Z

Zinc
  in acid copper-plating bath, 145
  EDTA and, 82
  in green beans, 128
  in plating solutions, 143, 144
$Zn^{2+}$, 78, 82
$Zn(CN)$, 90
Zn-EDTA, 84
Zwitterions, 15, 86, 115